Springer Theses

Recognizing Outstanding Ph.D. Research

Aims and Scope

The series "Springer Theses" brings together a selection of the very best Ph.D. theses from around the world and across the physical sciences. Nominated and endorsed by two recognized specialists, each published volume has been selected for its scientific excellence and the high impact of its contents for the pertinent field of research. For greater accessibility to non-specialists, the published versions include an extended introduction, as well as a foreword by the student's supervisor explaining the special relevance of the work for the field. As a whole, the series will provide a valuable resource both for newcomers to the research fields described, and for other scientists seeking detailed background information on special questions. Finally, it provides an accredited documentation of the valuable contributions made by today's younger generation of scientists.

Theses are accepted into the series by invited nomination only and must fulfill all of the following criteria

- They must be written in good English.
- The topic should fall within the confines of Chemistry, Physics, Earth Sciences, Engineering and related interdisciplinary fields such as Materials, Nanoscience, Chemical Engineering, Complex Systems and Biophysics.
- The work reported in the thesis must represent a significant scientific advance.
- If the thesis includes previously published material, permission to reproduce this must be gained from the respective copyright holder.
- They must have been examined and passed during the 12 months prior to nomination.
- Each thesis should include a foreword by the supervisor outlining the significance of its content.
- The theses should have a clearly defined structure including an introduction accessible to scientists not expert in that particular field.

More information about this series at http://www.springer.com/series/8790

Adhip Agarwala

Excursions in Ill-Condensed Quantum Matter

From Amorphous Topological Insulators to Fractional Spins

Doctoral Thesis accepted by
the Indian Institute of Science, Bangalore, India

 Springer

Author
Dr. Adhip Agarwala
International Centre for Theoretical Sciences
Tata Institute of Fundamental Research
Bangalore, India

Supervisor
Prof. Vijay B. Shenoy
Department of Physics
Indian Institute of Science
Bangalore, India

ISSN 2190-5053   ISSN 2190-5061   (electronic)
Springer Theses
ISBN 978-3-030-21513-2   ISBN 978-3-030-21511-8   (eBook)
https://doi.org/10.1007/978-3-030-21511-8

This Springer imprint is published by the registered company Springer Nature Switzerland AG
The registered company address is: Gewerbestrasse 11, 6330 Cham, Switzerland

"Why are you digging a hole?"

"I'm looking for buried treasure!"

"What have you found?"

"A few dirty rocks, a weird root, and some disgusting grubs."

"On your first try??"

"There's treasure everywhere!"

Bill Watterson, in Calvin and Hobbes (3rd June, 1995)

Supervisor's Foreword

Much of condensed matter physics focuses on crystals where there is a natural notion of periodicity. The idea that all is well understood from the point of view of electronic physics of crystals was washed away by the tsunami engendered by topological concepts.

Adhip Agarwala sets on a journey to explore "ill condensed matter", where there is no refuge of the Brillouin zone. In this thesis, he explores a variety of problems in amorphous systems, fractals, etc. with a view of exploring topological phases in them. He answers several interesting questions: Are topological electronic phases possible in amorphous systems? And, in fractal lattices? Read this thesis and you will be enlightened not only with answers to these questions but on how "fractional spins" can arise among other things.

The thesis also contains an excellent exposition of many of the modern concepts of condensed matter physics including topological ideas, fermionic symmetries, etc. I particularly invite younger scientists wishing to delve into the basics to enjoy this presentation by Adhip, one of my best teachers!

Bangalore, India
June 2019

Prof. Vijay B. Shenoy
Centre for Condensed Matter Theory
Department of Physics
Indian Institute of Science

Abstract

Impurities, disorder, or amorphous systems—the ill condensed matter—are mostly considered inconveniences to our study of materials, which is otherwise heavily based on idealized perfect crystals. Kondo effect and the scaling theory of localization are some of fundamental and early discoveries which had brought forth the novelty hidden in impure or disordered systems. Recent advances in condensed matter physics have emphasized the role of topology, spin-orbit coupling, and non-ordinary symmetries such as time reversal, charge conjugation, and sublattice in many physical phenomena. These have irreversibly transformed the essential ideas and purview of condensed matter physics, both in theoretical and experimental directions. However, much of these recent developments and their implications are limited to, or by ideas which stem from clean systems. This thesis deals with various aspects of these recent developments, but when they are unclean—introducing new ideas such as amorphous topological insulators, fractalized metals, and fractionalized spins.

Preface

Welcome!

Welcome to this thesis and to the stories presented in this. Curiosity often leads to terrains which are neither smooth nor clean—and therefore the name "Excursions in *ill* condensed matter". Disordered or *ill* systems have something beautifully mysterious about them. With their "faults, dislocations, defects" they show remarkably rich physics and a myriad of phenomena that have deep implications.

This thesis, as will be discussed at length in the first chapter, investigates recent developments in quantum condensed matter field such as topological phases, spin-orbit coupled systems, etc. in context of *ill* systems.

There are eight chapters. The first and the last serves as the introduction and the conclusion, respectively. The first chapter provides a broad overview of quantum condensed matter field and introduces the idea of *ill* condensed matter. It further places the motivations for various problems and sketches the central results. The second chapter discusses the formalism to classify quadratic Hamiltonians based on few intrinsic symmetries and shows how the generic structure of Hamiltonians can be obtained in each of the symmetry classes. Chapter three builds on these results and demonstrates the realization of amorphous topological insulators. Chapter four reports the surprises in realizing topological phases on fractals—systems which have a non-integer spatial dimension and where bulk and boundary is *ill*-defined. Chapter five takes another topological system—the Hofstadter butterfly—the spectrum when a square lattice is subjected to a magnetic field, and looks at the effect of randomly removing the lattice bonds. Chapter six explores impurities in a spin-orbit coupled system and shows the existence of fractional spins and a corresponding novel Kondo effect. Chapter seven extends the classification introduced in the second chapter, to arbitrary interacting systems and obtains the structure of many-body Hamiltonians in different symmetry classes. Chapter eight concludes with a perspective and discussion of future scope.

Most of the work presented in this thesis is already present in public domain, either through published manuscripts or arXiv preprints. Therefore, not unexpectedly, they will have close overlap with the content presented in them. Please feel free to get in touch for any clarifications or to send any suggestions, criticisms, etc.

Happy reading!

Bangalore, India   Adhip Agarwala

List of Publications

1. Adhip Agarwala, Arijit Haldar, and Vijay B. Shenoy
   The tenfold way redux: Fermionic systems with N-body interactions
   Ann. Phys. **385**, 469 (2017)

2. Adhip Agarwala and Vijay B. Shenoy
   Topological Insulators in Amorphous Systems
   Phys. Rev. Lett. **118**, 236402 (2017) (**Editor's Suggestion**) (Featured in
   Physics)

3. Adhip Agarwala
   Killing the Hofstadter butterfly, one bond at a time
   Eur. Phys. J. B **90**, 15 (2017)

4. Adhip Agarwala and Vijay B. Shenoy
   Quantum impurities develop fractional local moments in spin-orbit coupled
   systems
   Phys. Rev. B **93**, 241111 (2016) (**Rapid Communication**)

5. Adhip Agarwala, Shriya Pai and Vijay B. Shenoy
   Fractalized Metals
   arXiv:1803.01404

Acknowledgements

I think it was in 2011 when I first met Baskaran in Sankalpa's office at IIT Delhi. I had made up my mind to pursue a Ph.D. then. Baskaran inquired "where?" I had not decided yet. He, in his smiling self, told me "there is someone called Vijay Shenoy at IISc, have a look". The following summer, incidentally I was here for a fellowship through an Indian Academy of Science Program. On an afternoon I met Vijay and gave a talk in his group. And we met maybe another two times. There was no more turning back.

Vijay is an outstanding physicist, and his enthusiasm for physics is contagious. His simplicity, humbling. He surprises you with his creativity, with his ability to render most difficult of ideas in the simplest of words. His tenacity to continue with a problem and hitting it hard with all the strength. His flexibility to jump into unexplored areas and then make way through it. He challenges himself every day and challenges you by setting that example. It's rare to find a mid-aged man jump with excitement when he gets a new idea or understands something afresh; Or to come on a Sunday afternoon to the office, say, "Adhip! I have to tell you this!" and then follows an hour-long discourse on a solution which occurred to him during the morning stroll. In one of our first few meetings, even before I joined him for a Ph.D., Vijay had told me "Adhip, learn through discovery!" and if there is one thing I have seen him doing, any time and every time, is this—learning through discovery.

But that is just one part of Vijay. More than anything else he has been a friend. During the most difficult of times, Vijay stood as the strongest pillar of support. He has been a companion in grief, and happier than me in my happiness. He has shared my challenges and fought beside me. I can hardly thank him enough for everything so lets not even attempt to. Through Vijay, I also met Gopika Ma'am. I thank her for her concern and care; for the many breakfasts, lunches, park strolls, conversations, laughs, and giggles which I was fortunate to be a part of. And it has always been an absolute delight to watch Devanch grow up.

In these last 5 years, since I have come to know Vijay, it's difficult to count in how many ways he has influenced me. History is a set of random events. And given some of them had not happened, it's difficult to imagine the present. Coming across Vijay, was an accident. And I couldn't be more glad that it happened.

In this department I also met Prof. Diptiman Sen. He has had a deep impression on me, both physics-wise and in life, in general. In the last 2 years, we ended up working on multiple projects—and it has been an incredible learning experience. I thank him for all the patient hearing, for various chats and laughs, for teaching variety of things, for sharing his childlike interest in every discovery. His encouragement and support, his appreciation and care. I also thank Piyali ma'am for sharing some good appetizing times.

I acknowledge the quantum condensed matter theory group @ IISc. I have met a remarkable set of faculty members here. In addition to being remarkable physicists, they are remarkable individuals. In Manish, I have found a smiling friend, with whom I have discussed everything, be it physics or otherwise. I thank him for the many long conversations, for his concern and care. I will like to acknowledge Prof. Krishnamurthy, for his support at various junctures. As the chairman of the department, Prof. Krishnamurthy posed rare faith in us, when he allowed us to use lecture halls for student discussions, or to reopen the department library. Subroto, for physics discussions, and for his care and concern at numerous occasions. The complete cond-mat group in Bangalore, for many discussions, courses, journal club talks or seminars—Vijay, TVR, Diptiman, HRK, Subroto, Manish, Chandan, Rahul, Prabal, Tanmoy, Sumilan, Subhro, Samriddhi, Abhishek, Sanjib, Jainendra, Sriram. I also interacted with few experimental faculty members in the department —Arindam, Anindya and Aveek. I thank them for many discussions and collaborations.

"ehsaan tera hoga mujhpar..", one of the many songs Parveen Amma and I have sung together on the corridors. I will like to thank her for many scoldings, constant teasing, biryanis and many prayers she has said as we signed her cleaning records. With her I also acknowledge Reshma, Radha and other cleaning staff of our department for all their care.

I will now like to acknowledge the gangs, the goons and the pains of my life— my friends. It is impossible to acknowledge, the many terrible jokes, the loudly unmelodious songs, the crazy monkey dances, the late night food; Or the calm tea, the breezy walks on the roof, the many hundreds of conversations. In many ways, they have surrounded to shield, to support, to help, and when you least expect it—to pull your leg. I cannot express it enough, how lucky I am, to have met them, and to have spent this incredible time. Aabhaas. For sharing and being a partner in all our adventures, be it unjournal club, syahi, thex club, library; for being the pink dhoti wearing—weirdo, that he is; for being my last resort of all physics discussions, for many decisions he has helped me to take in my personal and academic life. Rituparno, for being equally insane; for hating things, which I hate. For feeding me all that food he can cook; for many things we have learned together. Parveen, for all the great times we have spent together, for his impatient listening to all my cribs and worries. Disha, for her constant unwavering support. Not only as she made every Mumbai trip more warmer, more welcoming and more easier for all of us, she and her crazy self has made many conversations a delight; to share, support, and help in millions of ways. Chandan, the same old "ghoda" that he has been since I had met him, 10 years ago in our college. He has been my support at many difficult times. Harsh, for being a brother. Debayan, for making us see the worst of movies, for speaking to us most inconsistent conversations and absurdest of lies. It's difficult imagining this time given he had not been a part of it. Oindrila, for never shying away from bringing me my flaws; for many sunset moments over the cup of lemon tea. Ranjani, for cracking equally terrible PJs; for the many songs we have sung on the department roof.

The early gangs: RPM who tried to instill all the civility in us, her scoldings, passion for food—both to gain and lose weight. Debuda, for his frog dances, anisotropic response to stimuli, for teaching me Fortran; Kinghsukda, Debshankar, Debadrita, Ananyoda, and Rituparno for creating some of the best moments in IISc. My office mates—Mit, my crazier neighbor, and partner in crime; for not losing any opportunity to make me feel a generation older! Jayanth, Arijit for ready help and many discussions. Sudeepda and Riadi, Akhilesh, Sudipta—for her various hallucinations, and the youngest member Kaushik.

My batchmates: Amit, who through his drama, Kishore songs, and poetry enriched our lives. Naren, for music and discussions; Soumen, Pari, for many great times; Deovrat, Kazi, Tirthankar, Avinash, Lokeshwar, Venkat, Anoop, Roobala, Sreekar, Suman, Gopal, Soumavo; My two "humsafars" from IIT Delhi—Pradip and Soling. Pradip, for leaving no opportunity to criticize my room or clothes. Many friends—Vipan, Sukanya, Ashutosh Bhaiya, Abhiram, Sambuddha, Umesh, Swati, Sauquibda, Aamir, Amogh, Sai, Bala, Rahul, Parnabda, Yogeshwar, Pramod, Sayonee, Ranjan, Rahman, Pushpendra, Satyendra, Kunjalata, Semonti, Anindita, Himangshu, Manisha, Debabrotoda, Atanu, Deya, Ankur. The presidency and associated gang, Rituparno, Debadrita, Saientan, Raj, Shantanu. Many younger friends: Shriya, Gopi, Saurav, Suman, Sangram, Prakriti, Hreedish, Kimberly, Gaurav, Sujay, Kaushik, Ramya, Soumi, Debasmita, Abhishek, Bhaskar, Saurabh, Deepak, Animesh, Surajit, Hemanta, Tathagata, Abhishek, Chandan, Mounika, Kingshuk, Mangesh, Krishna, Ravindra, Nimmi, Sukanya, Priyo, Mahesh, Indrajit, Aditya, Bhawana, Navneet, Nadia, Priya, Saloni, Ratan, Aradhita, Surabhi. Special mention of new gangs—Vikas, Moid, Ayush, Sudeb, the D-Mess gang (and NIAS tea gang) including Pandey, Aakaash, Anil, Rajiv, Babu, Dhanpal, and many others. The many members who make this department and make many wonderful moments possible here. For people in IISc, National law school, JNC, NIAS, and others, to whom I was connected through many causes. Geeta Ma'am, Ananth Sir, and associated people with Geetanjali.

My teachers Chaki Aunty, Abhay Sir, Pragati Ma'am, Sankalpa, Varsha, Dilip Ranganathan, Ajoy Ghatak, and Ajit Kumar. My super-seniors gang who are now in Bangalore—"Samriddhi, Subhro, Summilan", for all the leg pulling and concern. I acknowledge ICTS, its members, and people who made that institute possible. It has had innumerable contributions in many ways during these years. I will like to thank the present Chairman, Prof. Venkat. I am grateful to Srinivas, Srivatsa, Meena, Rekha, Bhargavi, Indu, Shanta, and others for helping with many official and technical things. Specially, Srivatsa for his care and concern. Mariappa,

Prabhakar for many help. The canteen downstairs—Ashwani and her family for keeping us refreshed. The institute, its admin, and its many supporting centers for making many things possible, and also for constantly reminding us of our vulnerabilities.

Finally, I will like to thank my family—the many people who define it, my first teachers—my parents, my sister, and my brother-in-law who all make everything worthwhile.

Bangalore, India   Adhip Agarwala
June 2017

Acknowledgement of Collaborative Contributions

1. The content presented in Chaps. 2 and 7 has equal contributions from Arijit Haldar. The collaboration is duly acknowledged.

2. The content presented in Chap. 4 has contributions from Shriya Pai. The collaboration is duly acknowledged.

Contents

Chapter 1
Introduction

1.1 Away from the Comfort Zone

Anderson in his famous paper announced "More Is Different" [1], a thought which he admittedly borrowed from Engels and Marx's three laws of dialectical materialism [2]; one of which is "The law of the passage of quantitative changes into qualitative changes". These ideas form the basis of the understanding and thinking about condensed matter systems, which have now been vetted through decades of careful theoretical investigations, experimental verifications and discoveries.

Heating a bowl of water may seem suspiciously innocent, until it suddenly decides to convert to steam. A heated magnet forgoes its love for a bar of iron at a specific temperature. These phenomena and many others are now weaved into a fascinating story of phase transitions, broken symmetry phases, Goldstone modes etc. now called the Landau paradigm. Phase transitions appear in two, surprisingly related, varieties—"classical" and "quantum". While our everyday lives can bring to us many a examples of classical phase transitions, i.e. ones executed via temperature, quantum phase transitions are harder to find. These typically happen at low temperatures, as a function of changing some other parameter, where quantum effects take the leading role. The exotic world of superfluids and superconductors belong to this territory. Their promise and scare to alter our world in incomprehensible ways is illustrated by the fact that when James Cameron imagined mankind's future in *Avatar*, he saw humans mining Pandora for *unobtainium*, a fictionalized room temperature superconductor [3].

The elaborate framework of phase transitions has been built on many exact and approximate theoretical studies on translationally invariant systems, or more colloquially, *clean systems*. Not surprisingly, condensed matter physicists are obsessed with clean systems. The concepts of a Fermi surface, Brillouin zone form fundamental processes of our thinking. But disorder is more than just an inconvenience. Little amounts of disorder is indeed tolerable in most situations, for example a block of copper continues to remain a metal, even with small aluminum impurities.

© Springer Nature Switzerland AG 2019
A. Agarwala, *Excursions in Ill-Condensed Quantum Matter*,
Springer Theses, https://doi.org/10.1007/978-3-030-21511-8_1

However, as eventually it became clear, often in systems even insignificantly small "quantitative" disorder can sometimes bring "qualitative" changes.

These discoveries surprised the community then and surprises every beginner now. Gang of Four's discovery that even tiniest amount of disorder on a two dimensional lattice can convert a metal into an insulator was absolutely unexpected [4]. So was the discovery that a few magnetic impurities can make an otherwise metal crossover to show insulating behavior at lower temperatures [5]. The discovery of both of these phenomena are considered landmarks in condensed matter physics and had implications much beyond. Both are also examples of systems where subtle effects of quantum mechanics wrestles to the forefront. Systems, even more drastically away from being lattice periodic are—amorphous, percolating and glassy. These have interesting phase transitions, crossovers and still have many fundamental open questions, even in the classical domain. The absence of any underlying lattice have posed significant challenges in building up the basic framework. Heavy numerical simulations and computational studies have provided some crucial phenomenology to uncover the physics. It is still a challenge before a complete underlying theory can be deciphered. Adding quantum mechanics is hardly expected to make things any simpler. While developments in theory of phase transitions and establishment of Landau paradigm had brought some semblance of understanding phase transitions etc., it is still quite awkward to think about these systems in terms of symmetry broken phases. The exhaustion to deal with these may be sensed by the little fact that, when a school was conducted in 1978 by stalwarts in the field including Anderson, Thouless, Kirkpatrick, Lubensky—where many of these systems were discussed—they found no phrase better than "*Ill* condensed matter" to name this school [6].

1.2 Something Topologically Different

K.G. Wilson was awarded the 1982 Nobel prize in Physics for "his theory for critical phenomena in connection with phase transitions". It was the period when the theory of phase transitions and its analysis using renormalization group had taken deep roots. Its successes in explaining "universality", where microscopically different systems showed identically similar properties at critical points during phase transitions, was phenomenal [7]. Not unexpectedly, "Landau–Ginzburg–Wilson" was the paradigm now. Ironically, just two years before, 1980 was also the year, first observations of quantized Hall effect was reported [8]. In 1985, this discovery was awarded the Nobel Prize [9].

The germs of a new physics was now in the open, and that it had opened a Pandora's box can be realized by the fact that the 2016 Nobel prize in Physics was quite unexpectedly[1] given "for theoretical discoveries of topological phase transitions and topological phases of matter".

---

[1]Unexpected, since the race included LIGO's discovery of gravitational waves! [10].

Between 1980–2016, in these intervening years, quantum condensed matter has seen a major makeover. Topological systems have made a startling appearance and are now an essential part of the condensed matter diet. That the Hall physics has topological connections was identified early [11]. But slowly the ideas burgeoned into identifying new materials and new phenomena. As we shall see in the next section, there is something strangely different in quantum Hall physics, and many others which have followed its footsteps. Today these phases of matter, quite literally, are *topologically different*.

1.3 Phases and Symmetries

The phases of matter which fall in Landau–Ginzburg–Wilson paradigm or are "topological" in nature have one thing in common—their reverence for symmetries. Symmetries plays an essential role, in both realization and identification of these various phases, as well as to analyze the phase transitions.

An object is called symmetric, even colloquially, when it continues to look the same even if some transformation is done to it (see Fig. 1.1). The same applies to Hamiltonians. If a symmetry operation commutes with a Hamiltonian we say that the system contains that symmetry. Landau-like phase transitions are accompanied by "breakage" of any parent symmetry. A canonical example being a set of N classical ising spins in two dimensional square lattice, where neighboring spins interact through a ferromagnetic coupling $-J$ ($J > 0$); and one wishes to investigate the magnetization as a function of temperature. The Hamiltonian is given by

$$\mathcal{H} = -\sum_{\langle i,j \rangle}^{N} J S_i^z S_j^z \qquad (1.1)$$

where $\langle i, j \rangle$ denote the nearest neighbor spins. The $S_z \to -S_z$ symmetry in the Hamiltonian gets broken in the ground state, when either all the spins are pointing up, or are pointing down. This occurs at a particular temperature given by the critical temperature T_c (see Fig. 1.1) . This explains why a bar of magnet suddenly becomes non-magnetic at a particular temperature.

More generically, depending on the symmetries present in a system, one can write the lagrangians in terms of the symmetry allowed terms, and analyze the various phases as an interplay of preferred ground states with original symmetries broken. Further, various phase transitions and their critical properties become independent of the microscopic details—bringing in an utopian universality among various phases.

Topological phases respect symmetries in a different way. Away from instincts of breaking a symmetry, topological phase transitions are quite protective about it. Therefore these phase transitions occur without a symmetry change.

But for starters, what are topological phases and their transitions? The case in point is the quantum Hall curve as shown in Fig. 1.2. The setup comprises of a two

 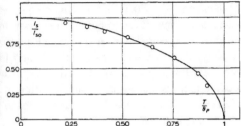

Fig. 1.1 Symmetries and phases: (Left) The road side tea in India is typically served in a glass such as shown above. It is also symmetric given a reflection about a central plane as denoted by the dashed line. (Source:Internet) (Right) Temperature dependence of the spontaneous magnetization of magnetite (circles are experimental points and the curve is obtained within the mean-field theory.) Adapted from the Nobel lecture of Neel [12]

dimensional electron gas, subject to a perpendicular magnetic field. Landau showed as early as in 1930 [13] that this system has an infinite spectrum of equally gapped flat bands. Therefore with some stroke of luck, if the Fermi energy lies between the gaps, one should have an insulator. This implies that any measured conductance between the two leads should show *zero*. The experimentally found curve is also shown in Fig. 1.2. The measurement of ρ_{xx} is zero, in most region, except at isolated points. Quite surprisingly, ρ_{xy} shows plateaus which are quantized at $\frac{1}{n}\frac{h}{e^2}$—a number only dependent on fundamental constants. It was also understood that the current, which contributes to this conductance moves only through the edges of the sample. This phenomena has three crucial features, also shared by most topological phases: First, this otherwise was supposed to be an insulator; Second, the transport happens through the edges; Third, the conductance is precisely quantized i.e., it does not change for different kinds of samples etc. These features were understood by the fact that the band gaps in quantum Hall problem are not *trivial* in nature, but instead *topological*.

A generic topological gap, mostly, share properties such as edge transport, quantized conductance and some robustness—such as the inability of disorder to disrupt the conductance from a quantized value. Topological band gaps, like in the case of quantum Hall, can be differentiated from trivial band gaps through some topological number, which takes nonzero integral values in the topological phases (see Fig. 1.3). Robustness to disorder is in fact mapped to inability of changing this topological number in a smooth fashion. In systems, one can go from a topological gap to a trivial gap, through a phase transition—called a topological phase transition. At this transition the topological number characterizing the phase, changes its value. n, which is also the value of the quantized transverse conductivity in quantum Hall is one such number. To illustrate this, we consider a simple topological model in one dimension [14]—a set of fermions, denoted by creation and annihilation operators at every site i with $c_i^\dagger$ and c_i, hopping on the lattice of equally spaced N sites, respecting the following Hamiltonian

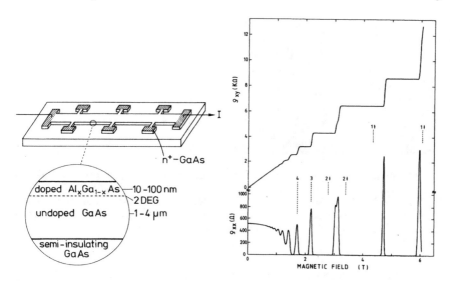

Fig. 1.2   Integer quantum Hall effect. (Left) The Hall setup of Klitzing and collaborators for a Hall measurement on two dimensional electron gas. (Right) The measured low temperature transverse and longitudinal resistivity. Figures adapted from the Nobel lecture of Klitzing [9]

$$\mathcal{H} = \sum_{i=1}^{N} t_1(c_i^\dagger c_{i+1} + h.c) + t_2(c_i^\dagger c_{i+2} + h.c.). \tag{1.2}$$

For a density of fermions which is one electron for every two sites, this system is an insulator at any value of $|t_2| \neq t_1$. However when $|t_2| < t_1$, an open chain of this model contains a pair of edge states, not unlike the current carrying edge states of the quantum Hall. This phase is topological in nature and is in fact different from a trivial insulator ($|t_2| > t_1$) due to an integer, called the winding number. This is shown in Fig. 1.3. $|t_2| = t_1$ signifies the topological phase transition, and unlike the magnet we discussed before, no symmetry gets broken here. In fact, as we will discuss later, some symmetries need to be preserved for this transition to occur.

With discovery of topological phases and its identification in quantum Hall, the community started to realize and seek similar physics in various other systems. Various other models and phenomena, were now identified as 'topological', be it in polymer chains [14] (this is the model we discussed above), in a model without magnetic field [15], in graphene [16, 17], in Majorana chains [18] etc. All these transitions occur without breaking any symmetry.

It slowly emerged that all these topological phases fall in a pattern and are in fact dependent on *few* underlying symmetries. However, these symmetries are quite out of the *ordinary*.

Time reversal symmetry, charge conjugation symmetry, and sublattice symmetry are three non-ordinary symmetries, a topic which will be discussed in much detail in this thesis. Based on these three, *all* fermionic systems can be classified into ten

 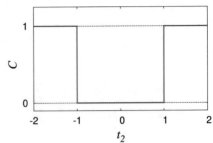

Fig. 1.3 Topology and phases: (Left) The south Indian delicacies, idli (white ellipsoid) and vada (brown torus), are topologically distinct due to the number of punctures through them. This "number of punctures" is a topological number. (Source:Internet) (Right) The topological phase transition for the model described in text (see Eq. (1.2) occurs at $t_2 = \pm 1$. The integer (C) characterizing the topological phase is similar to the number of punctures in a idli/vada

classes. In 1997, Altland and Zirnbauer had showed this classification [19]. It turns out that for topological transitions, for a Hamiltonian belonging to one of the ten classes, these symmetries need to be kept intact. By 2009, classification of *all* topological phases based on these ten classes and spatial dimensions was discovered [20, 21]. Kitaev in [20] answered two important questions: First, given a set of symmetries and a spatial dimension—can a system host topological phase? Second, if yes, what will be the nature of the "topological number"?

The episode of topological physics is a remarkable advancement in saga of condensed matter physics. It has enriched and built on what the Landau–Ginzburg–Wilson paradigm had led us to. Another remarkable achievement of the field in the last few years, have been to realize the importance of these non-ordinary symmetries in condensed matter physics. Many of these ideas will be discussed in this thesis. In the next section we embark upon revisiting some of the central milestones in condensed matter physics. These are also intended to whet the appetite for what will be presented in the forthcoming chapters of this thesis.

1.4   The Landmarks: The *Essential* Milestones

1.4.1   *Anderson Localization and Scaling Theory of Localization*

One of the first examples taught to us in any quantum mechanics course is that of a "a particle in a box". This introductory example manifests a fundamental behavior of quantum particles, and therefore electrons. Energy levels of a particle in a (d-dimensional) box of volume L^d is given by (Fig. 1.4)

Fig. 1.4 Extended and localized states: (Left) Typical wave function of an extended state with mean free path l, and (Right) a typical wave function of a localized state with a localization length ξ. Adapted from [25]

$$E_{\{n\}} \propto \sum_{i=1}^{d} \frac{n_i^2}{L^2} \tag{1.3}$$

where $\{n_i\}$ is a set of d integers $\in (0, \ldots, L-1)$ (which will serve as a quantum number) and L is the length of any side of the box. Confining a particle costs energy. Therefore, given a choice, any quantum particle would like to spread itself in the whole space. The eigenstates of the system can be identified with a discrete set of wavelengths and therefore a momentum $\boldsymbol{k} = \{k_i\} \equiv \{n_i\}\frac{2\pi}{L}$. This discrete set of points form a hypercubic lattice in d dimensional momentum space. The simplest "spherical cow" model of any material can be assumed to be a finite density of noninteracting spinless fermions residing in this d-dimensional box. Being fermions, in pursuit of minimizing energy (even at temperature $T = 0$), what all electrons can do at best is to occupy some set of states upto a finite energy called the Fermi energy. In $d = 1, 2$ or 3, the Fermi surface is a set of two points, a circle or a sphere respectively and there are similar "lattice-based" tight-binding analogs of free fermionic systems. These free fermion models explain many of the essential characteristics of metals, be it their electronic specific heat, magnetic susceptibility or the fact that they are gapless to excitations i.e., they can pass current.

Anderson in [22] showed that, under certain conditions, electrons can decide to remain localized to a region, rather than spread out over the whole lattice. Some excellent reviews on physics of this can be found in [23–26] and we will discuss the essential ideas here. The Anderson model is described by a set of fermions hopping on a lattice whose onsite (ϵ_i) energies are disordered. The Hamiltonian is given by

$$\mathcal{H} = -\sum_{\langle i,j \rangle} t_{ij}(c_i^\dagger c_j + h.c) + \epsilon_i n_i \tag{1.4}$$

where ϵ_i are uncorrelated and drawn from a box potential $[-\frac{W}{2}, \frac{W}{2}]$. Here W characterizes the strength of disorder. Anderson argued that an electron in such a system can remain localized in a region of space with a wave function of the following form

$$|\psi(\boldsymbol{r})| \sim \exp(-(\boldsymbol{r} - \boldsymbol{r}_o)/\xi) \tag{1.5}$$

where ξ is the localization length, which characterizes the length over which the wave function can be considered to be localized, and r_o is the centre of the localized state. ξ has a dependence on W which we will discuss later. A transparent way to see this physics is by the following arguments (this follows closely the discussion by Ramakrishnan (see [27] and references therein)). Onsite Green's function for a decoupled site i, with onsite energy ϵ_i is given by

$$G_{ii}(E + i\eta) = \frac{1}{E + i\eta - \epsilon_i} \tag{1.6}$$

where $E^+ \equiv E + i\eta$, where $\eta \to 0$, and $\eta > 0$. $-\frac{1}{\pi} \operatorname{Im} G_{ii}$ is called the spectral function, which crudely is the probability of finding a particle at this site at some energy E. Since this decoupled site i has an onsite energy ϵ_i, the spectral function will be a δ function centered at energy ϵ_i. In presence of hopping to other sites, G_{ii} gets modified to

$$G_{ii}(E^+) = \frac{1}{E^+ - \epsilon_i - \Sigma(E^+)} \tag{1.7}$$

where, $\Sigma(E^+)$ is the self-energy. If $\Sigma(E^+)$ is such that $\sim \operatorname{Im} G_{ii}$ has a finite value (even as $\eta \to 0$) at all energies, one can safely consider the system to be delocalized since, at any energy, one can have a finite probability of finding a particle at site i. On the other hand, if $\sim \operatorname{Im} G_{ii} \propto \eta$ as $\eta \to 0$, this implies that we still have a δ function like spectral function, implying a localized state. This follows from the observation that there exists energy ranges in the spectrum, where one cannot populate the particular site i. Therefore properties of $\Sigma(E^+)$ can inform whether a spectrum is localized or delocalized. $\Sigma(E^+)$, in the approximation of small hopping integral, is given by

$$\Sigma(E^+) = \sum_{k \neq j} \frac{t_{ij} t_{jk}}{E^+ - \epsilon_k}. \tag{1.8}$$

As has been detailed in [27], the average of $\langle \operatorname{Im} \Sigma(E^+) \rangle$, is indeed a finite quantity as $\eta \to 0$. This *incorrectly* suggests, that the system has a delocalized spectrum. What one needs to look at is the distribution of $\operatorname{Im} \Sigma(E^+)$ where one finds that probability of obtaining such finite $\operatorname{Im} \Sigma(E^+)$ (as $\eta \to 0$) is vanishingly small, thereby implying *localized states*. Here one finds an interesting distinction, where physics is not determined by the average of a quantity but rather its probability distribution. What these arguments illustrate is the case where a quantum mechanical particle may chose to stay localized to a region in space rather than spread out to the complete lattice.

Given a band of states, one can further argue (i) that due to disorder, such localized states will first appear at the edge of the band, (ii) At a particular energy, both localized and delocalized states cannot co-exist. A "mobility edge" (in energy $\equiv$ E_c), therefore, distinguishes between such states. The localization length which we mentioned earlier (ξ) is finite in the localized regime and diverges when E approaches E_c. With increasing W the mobility edge moves towards center of the band and at

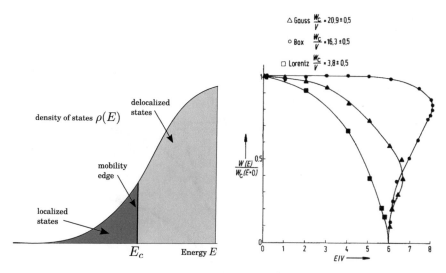

Fig. 1.5 Mobility edges: (Left) The mobility edge, E_c, separates the localized states in the band tail from the conducting states. By changing the Fermi energy, E_F, the system exhibits a transition from the metallic regime ($E_F > E_c$) into the localized regime ($E_F < E_c$). (Right) Numerically calculated mobility edge trajectories as a function of increasing disorder for three types of disorder kinds in three dimensional Anderson insulator model. Adapted from [28, 29]. As W increases, E_c reaches $E = 0$ signifying that all states are localized

some critical $W = W_c$ all states will get localized. This analysis in fact matches well with the numerical analysis of Anderson localization (see [28] and references therein; also see Fig. 1.5).

One can therefore imagine a situation where the Fermi energy lies in the a band of localized states, due to which the system is a transport insulator. Note that this is fundamentally different from a band insulator where the insulating behavior occurs because the Fermi energy lies in a band gap (i.e., density of states $\rho(E_F) = 0$). One of the questions which arises is how W_c depends on spatial dimension. In $d = 1$ it was known that all states are localized by any amount of disorder. The main point of contention was $d = 2$.

A remarkable achievement in this direction was made by the landmark paper of Gang of Four [4], which through renormalization group arguments showed that any infinitesimal disorder will localize the electrons in two spatial dimensions. This gave the single-parameter scaling theory of localization which we now discuss [4, 25]. The essential idea is to recognize that conductance is the only parameter in the theory and as one scales the system size, the changed value of conductance should *only* depend on the value of conductance prior to scaling the system size. At larger value of conductance where we have a diffusive metal, Ohm's law should be valid i.e.,

$$\sigma = \sigma_o L^{d-2}. \tag{1.9}$$

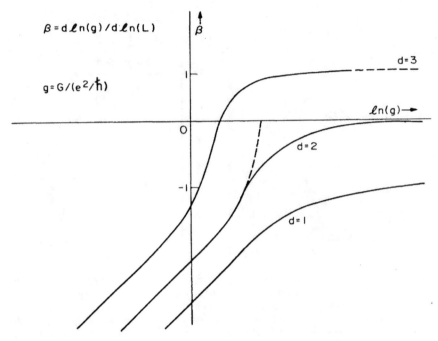

Fig. 1.6 The Gang of Four scaling theory: The behavior of $\beta = \frac{d \log g}{d \log L}$ as a function of $\log g$. Adapted from [25]

While we know in a Anderson localized regime, conductance will fall exponentially with system size i.e.,

$$\sigma = \tilde{\sigma}_o \exp(-L/\xi) \tag{1.10}$$

The next step is to identify the crucial parameter $\beta = \frac{d \log g}{d \log L}$ and look at its behavior with g. g is the dimensionless conductance $\equiv \sigma/(e^2/h)$ and was earlier recognized as an important quantity by Thouless and collaborators, who related it to the effect of boundary conditions on the wave functions (see [24] and references therein). At large g, $\beta \to d - 2$. While at small g, $\beta \sim \log g$ which will be a large negative quantity with decreasing g. Under the assumption of continuity, one can trace out the complete behavior of β and is shown in Fig. 1.6.

The important point to note is the sign of β for different dimensions. In $d = 3$, β switches sign at a certain value of $g \equiv g_c$. $\beta > 0$ for $g > g_c$ and $\beta < 0$ for $g < g_c$. This implies that with increasing L, g will increase if initially $g > g_c$ and will decrease when $g < g_c$. These two distinct behaviors imply existence of a metal insulator transition in $d = 3$. However given $\beta < 0$ in the complete regime for $d \leq 2$, one concludes that one will always find a insulating behavior. In fact, perturbative calculations [25] show that $\beta(g)$ goes as

$$\beta(g) = d - 2 - \frac{a}{g}. \tag{1.11}$$

This, therefore established that in two spatial dimensions, *any* disorder should be able to localize a thermodynamic two dimensional metal. A detailed discussion of this exciting field and the innumerable contributions of various people is beyond the scope of this few introductory pages. Interested reader may find discussions regarding the essential ideas and theory in [24–27] and some detailed numerical studies given in [26, 28] useful.

1.4.2   Local Moments and Kondo Effect

In the previous section, we visited the ideas of localization where random onsite potentials can localize an electron. The analysis and conclusions there are all for noninteracting systems i.e., the Hamiltonian is quadratic in form. What happens if we have an *interacting* impurity in a metal? This question is crucial and has implications in formation of local moments, magnetism, strongly correlated materials etc. We will visit the essential ideas of this theory in this section.

The prototype model was introduced by Anderson in [30], now called the Anderson impurity model and has the following ingredients: A free fermion metal also called the bath is made by states $c_{k\sigma}^{\dagger}$ with a dispersion ϵ_k; an impurity site d with onsite energy ϵ_d, and a hybridization through a potential V between the impurity and the metal. The Hamiltonian is

$$\mathcal{H} = \sum_{k\sigma}(\epsilon_k - \mu)c_{k\sigma}^{\dagger}c_{k\sigma} + \sum_{\sigma}\epsilon_d d_{\sigma}^{\dagger}d_{\sigma} + U\hat{n}_{d\uparrow}\hat{n}_{d\downarrow} + \sum_{k,\sigma}\left(\frac{V}{\sqrt{\Omega}}d_{\sigma}^{\dagger}c_{k\sigma} + \frac{V^*}{\sqrt{\Omega}}c_{k\sigma}^{\dagger}d_{\sigma}\right).$$
$$\tag{1.12}$$

In the limit of $U = 0$, one can evaluate the impurity Green's function which is given by

$$\mathcal{G}_{d\sigma,d\sigma}(E^+) = \frac{1}{(E^+ - \epsilon_d - \sum_k \frac{V^2}{\Omega}\frac{1}{(E^+ - \epsilon_k)})}. \tag{1.13}$$

The spectral function $\rho_{d\sigma}(E) \equiv \frac{-1}{\pi}\text{Im}(\mathcal{G}_{d\sigma,d\sigma})$, as we had seen in the last section, signifies the probability of populating the impurity site at energy E. One can also calculate the occupancy of the site by (at temperature $T = 0$),

$$\langle n_{d\sigma}\rangle = \int_{-\infty}^{\mu}\rho_{d\sigma}(E)dE \tag{1.14}$$

where μ is the chemical potential. While this is straight forward to calculate in the case of a noninteracting problem, one can extend this analysis to the interacting case under some assumptions. Anderson performed a mean field analysis of this model where the original interaction term in the Hamiltonian is replaced by,

$$U \hat{n}_{d\uparrow} \hat{n}_{d\downarrow} \rightarrow U \left(\hat{n}_{d\uparrow} \langle \hat{n}_{d\downarrow} \rangle + \langle \hat{n}_{d\uparrow} \rangle \hat{n}_{d\downarrow} - \langle \hat{n}_{d\uparrow} \rangle \langle \hat{n}_{d\downarrow} \rangle \right) \qquad (1.15)$$

One now evaluates $\langle n_{d\sigma} \rangle$ self consistently (the details can be found in [30, 31]) and find $\langle n_{d\uparrow} - n_{d\downarrow} \rangle$, the result is shown in Fig. 1.7. This quantity signifies if a magnetic moment is formed at the impurity site. Anderson assumed a flat band density of states for the metal with a constant value ρ_o and kept V perturbative. The essential result is that there exists a finite region in the phase space where $\langle n_{d\uparrow} - n_{d\downarrow} \rangle$ is nonzero.

It is particularly interesting to look at the decoupled limit, i.e., when $V = 0$, also called the "atomic limit". While a singly occupied d site has a gain in energy of $\mu - \epsilon_d$, a doubly occupied d site will have to pay additional energy of U. Now consider $\epsilon_d < \mu$, which means that it is favorable to occupy a site at the noninteracting level. Two electrons, one at the Fermi energy and the other at the impurity level has therefore the total initial energy of $\epsilon_d + \mu$. When the other electron decides to shift to the impurity site, the energy of this configuration is $2\epsilon_d + U$. Now if the final configuration is higher in energy than the previous state, we expect a local moment. This condition implies,

$$2\epsilon_d + U > \epsilon_d + \mu \rightarrow \frac{\mu - \epsilon_d}{U} < 1 \qquad (1.16)$$

Also if $\mu - \epsilon_d < 0$ the impurity level is above the Fermi energy, and therefore it won't be favorable to occupy it at all. Thus in the decoupled limit, it is favorable to form a local moment when $0 < \frac{\mu - \epsilon_d}{U} < 1$. This explains the local moment regime in the vertical axis of Fig. 1.7. Interestingly, if one is in the local moment regime, and increases the density of states of the hybridizing metal, one will eventually lose the magnetic moment. This basic model therefore offers a microscopic mechanism of a local moment formation.

The discussion above suggests that if in an experiment we reduce the temperature below the U scale, a local moment will be formed at the impurity site and continue to remain so till $T = 0$. However, this was eventually realized to be incorrect. While a spin is indeed formed at the impurity, it can still interact with the electrons of the metal to form a singlet state. The low energy model in this limit, called the sd model, is given by

$$H = \sum_{k\sigma} \epsilon(k) c_{k\sigma}^\dagger c_{k\sigma} + J \boldsymbol{s} \cdot \boldsymbol{S}. \qquad (1.17)$$

Here $\boldsymbol{S}$ is the spin at the impurity site while $\boldsymbol{s} \sim \sum_{k\sigma, k\sigma'} c_{k\sigma}^\dagger \boldsymbol{\tau}_{\sigma,\sigma'} c_{k'\sigma'}$ is the spin formed by the bath electrons. Jun Kondo in [5] showed that due to spin-spin scattering there would be a logarithmic increase in resistance with reducing temperature. This in fact was seen in the experiments [32] (see Fig. 1.7). The antiferromagnetic J scale appearing here can be related to the parameters in the Anderson impurity model (see Eq. (1.12)) as $J \sim \frac{V^2}{U}$ [31]. The intuitive idea to understand this term is that the impurity site can be virtually occupied by both the spins by paying energy cost U, but allowing for a spin exchange process. While Jun Kondo's analysis showed that

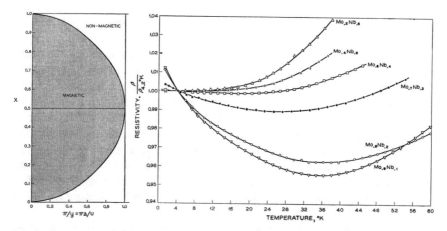

Fig. 1.7 The Kondo effect: (Left)The plot of $\langle n_{d\uparrow} - n_{d\downarrow} \rangle$ as a function of $x = \frac{\mu - \epsilon_d}{U}$ and $\frac{\pi \Delta}{U} = \frac{\pi \rho_0}{U}$ (see text). Adapted from [30]. (Right) Resistivity vs temperature for various Mo-Nb alloys containing 1% Fe. A rise in resistivity with reducing temperature can be seen. Adapted from [32]

reducing temperature produce an increase in resistivity. What happens at $T \to 0$? Kondo's calculation predicts that as $T \to 0$ the resistivity diverges. The calculation further predicts that susceptibility also diverges. The analysis in fact breaks down at a nonperturbative temperature scale T_K called the Kondo temperature,

$$T_K = De^{-\frac{1}{2J\rho_0}}. \tag{1.18}$$

Here D is bandwidth of the bath states. What was clear is that the the perturbative scattering calculation was not valid as T approaches T_K. A way to approach this problem was therefore by setting $T = 0$ and explicitly solving for the ground state. A initial development in this direction was a variational calculation developed in [33]. Here one considers a rigid Fermi sea, i.e., one assumes that the Fermi sea does not directly participate in the physics, but one considers all the two-particle states which can be possibly constructed with the impurity site and the states of the metal. Yoshida chose variational ansatzes of the form [31],

$$\Psi_{s,t} = \sum_{k > k_F} \alpha_k (c^\dagger_{k,\downarrow} d^\dagger_\uparrow \mp c^\dagger_{k,\uparrow} d^\dagger_\downarrow) |\Omega\rangle \tag{1.19}$$

where $|\Omega\rangle$ is the filled Fermi sea. s, t denote the singlet and triplet state. The variational calculation showed that a singlet ground state is preferred over a triplet ground state. The energy gain is of the order of T_K. More sophisticated variational ansatz were further employed which lead to same qualitative conclusion (see [31] and references there in). This therefore leads to the conclusion that at $T = 0$ the system is *non-magnetic*. While results of Anderson impurity model suggests that a local moment will be formed—it is now understood that with reducing temperature, this local moment will get *screened*.

A complete picture was derived by implementing ideas of renormalization group in Kondo problem [7, 34] and development of numerical renormalization group (NRG) by Krishnamurthy, Wilkins and Wilson [35, 36]. The essential idea in Anderson's poor man scaling [34] was to scale the bandwidth of the bath states, and ask how the antiferromagnetic coupling J scales with it. It was shown that reducing the bandwidth makes the effective J stronger and stronger. Numerical implementation of these ideas was done by mapping the full Kondo problem into a one-dimensional problem by a logarithmic discretization of the bath states. A just treatment of this fantastic theory is beyond the scope of this introductory section and the reader is suggested to refer to the original papers [35, 36]. We, however, discuss the essential results.

One of the central results of these works is the plot of susceptibility and temperature as shown in Fig. 1.8 for the Kondo model. The important point to note is that when $T \gg T_K$, $\chi T \to$ constant, implying a Curie law behavior. This is the manifestation of the local moment regime. However, as $T \to 0$, one notices that $\chi T \to 0$, meaning that susceptibility becomes independent of temperature. This is a trademark behavior of a spin-singlet metallic phase. This is the statement that the local moment for $T < T_K$ gets screened by the bath electrons, and thereby the spin can no longer be "seen". This many body singlet state comprising of the impurity and the associated bath states is called the Kondo singlet.

The above analysis is true in the Kondo regime, where the spin at the impurity is well formed, i.e., $U \gg T$. It is interesting to understand the complete picture, in presence of U and starting from the Anderson impurity model. Again it is instructive to analyze the atomic limit, i.e., when the impurity is decoupled to the bath. The energy scales in the problem are U, ϵ_d and T. The impurity site has four possible states: (i) unoccupied, (ii) and (iii) singly occupied with either spin, (iv) doubly occupied. We now look at the expectation of spin-moment $\langle (n_{d\uparrow} - n_{d\downarrow})^2 \rangle$ in various temperature regimes and since $\chi T = \langle (n_{d\uparrow} - n_{d\downarrow})^2 \rangle$ we expect susceptibility to take these various values as a function of temperature in the decoupled limit. The expression is given by,

$$\chi T = \langle (n_{d\uparrow} - n_{d\downarrow})^2 \rangle = \frac{2e^{-\beta\epsilon_d}}{1 + 2e^{-\beta\epsilon_d} + e^{-\beta(2\epsilon_d + U)}}. \tag{1.20}$$

It is important to note that we are working in units where a single electron has a unit local moment. When $T \gg U$ all the states are equally likely and therefore $\chi T = \frac{1}{2}$. This is called the free-orbital (FO) regime. When $T \ll U$, such that the impurity is singly occupied $\chi T \to 1$, this is called the local-moment (LM) regime. In presence of hybridization to the metal, it is in the local moment regime, that when $T < T_K$, χT starts to deviate and reach zero showing that the local moment has been screened. The free-orbital and the local moment regimes are the only two regimes possible if we allow that the impurity site either prefers to be singly occupied ($T \ll U$), or has equal likelihood for all the four states ($T \gg U$). This assumes that, in the local

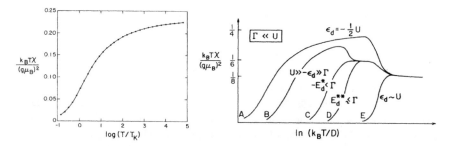

Fig. 1.8 Kondo signatures: (Left) Universal plot of susceptibility with temperature for the spin-$\frac{1}{2}$ Kondo Hamiltonian. (Right) The different values of χT which are reached as a function of temperature T for different values of ϵ_d (Γ here represents the hybridization). Adapted from [36]

moment regime, the system equally dislikes the state of being unoccupied at all (state (i)) or being doubly occupied (state (iv)). This assumption implies $2\epsilon_d + U = 0$, or $\epsilon_d = -\frac{U}{2}$. If ϵ_d is set to this particular value, the model is called the *symmetric* Anderson model. However, in general ϵ_d can be different from this value. This is called the *asymmetric* Anderson model. In asymmetric Anderson model, one has another regime apart from FO and LM. Here when $\epsilon_d < T < U$ the system may prefer single occupancy and no occupancy. In this case $\chi T = \frac{2}{3}$. This is called the asymmetric local moment (ALM) regime. As the discussion of the atomic limit clarifies these fixed points, in the presence of hybridization to the bath, the system will reach another qualitatively distinct fixed point. This is the Kondo singlet fixed point, where when as $T < T_K$, $\chi T \to 0$. The $\langle (n_{d\uparrow} - n_{d\downarrow})^2 \rangle$ still remains 1 in the Kondo singlet regime, but the susceptibility becomes independent of temperature proving the freezing of the impurity spin. These regimes can be seen in Fig. 1.8, which has been reproduced from [36]. The curves of interest are A and B. A shows the behavior for the symmetric Anderson impurity model and B represents that of asymmetric Anderson model. The values shown in this plot can be obtained by multiplying $\frac{1}{4}$ times the values we have discussed (these works choose spin moment to be 1/2 , hence the values should be compared to $\frac{1}{4}\langle (n_{d\uparrow} - n_{d\downarrow})^2 \rangle$). While both A and B curves start from the FO regime at large T and reach LM at low temperatures, the curve B visits another intermediate plateau which is the ALM fixed point. Finally at much lower temperatures both reach the Kondo regime as $\chi T \to 0$. This physics, apart from the NRG methods, have also been verified by development of quantum Monte Carlo methods [37].

This concludes our brief discussion on the Kondo effect. Many of these ideas will be revisited in a later chapter. Interested readers are urged to visit the original articles [31, 35, 36] for many details and insights.

1.4.3   *Topological Band Theory and Topological Invariants*

In the last two subsections we have visited the ideas of Anderson localization and then Kondo effect. We now discuss a later development in the condensed matter physics arena which is that of topological band theory. In Sect. 1.3 we briefly discussed integer quantum Hall physics and the Su-Schrieffer-Heeger (SSH) model. We showed that for the SSH model one in fact has a topological number which changes discontinuously as a function of t_2. We now discuss the essential ideas behind these kinds of topological numbers, ideas of bulk-edge correspondence and meaning of topological protection to scattering.

Before anything else the central concepts on which topological band ideas are based on are adiabaticity and Pancharatnam–Berry phase which we now discuss [38, 39]. Say, a parameter in the Hamiltonian is changed from one value to another— suddenly. In that case the ground state of the initial Hamiltonian is unable to adjust to the new Hamiltonian and it goes into excited states of the final Hamiltonian. However, it might be possible for a Hamiltonian to be smoothly changed such that the wave function changes smoothly going from ground state of the initial one to the ground state of the other. However in this process a nontrivial phase may be acquired which is called the Pancharatnam–Berry phase. Consider a Hamiltonian which depends on some parameter R and has the ground state energy $E_g(R)$ and corresponding ground state wave function as $|\psi(R)\rangle$. Now the parameter R is changed from an initial R_{in} to final R_{fi}. Pancharatnam–Berry phase is then given by,

$$\gamma = i \int_{R_{in}}^{R_{fi}} \langle \psi(R)| \frac{d}{dR} |\psi(R)\rangle dR \qquad (1.21)$$

R_{fi} can be same as R_{in}, which will imply a closed path. If R is a scalar variable then an integer derived from this expression is called the winding number. However, in general R can be a vector. Then there is an interesting interpretation to the Pancharatnam–Berry phase. One can associate $\langle \psi(\mathbf{R})| \frac{d}{d\mathbf{R}} |\psi(\mathbf{R})\rangle \equiv A(\mathbf{R})$ which is like a "vector potential". One can also associate a corresponding magnetic field called the Berry curvature

$$F(\mathbf{R}) = \nabla_{\mathbf{R}} \times A(\mathbf{R}). \qquad (1.22)$$

A surface integral of the Berry curvature will also lead to the Pancharatnam–Berry phase. One of the paradigmatic examples to see the nontrivial effects of such a phase is to study the effect of a rotating magnetic field on a spin-$\frac{1}{2}$ object. Considering the Hamiltonian to be $-\boldsymbol{\sigma} . \boldsymbol{B}$ where $\boldsymbol{B}$ is the magnetic field $\boldsymbol{B} = (B_x, B_y, B_z) = B(\sin\theta \cos\phi, \sin\theta \sin\phi, \cos\theta)$ and the spin points in the direction of this magnetic field. If we rotate $\boldsymbol{B}$ about the z axis, and reach the same point the wave function acquires a phase of $-2\pi \sin^2(\frac{\theta}{2})$. The other wave function takes the same phase with an opposite sign. Consider the special case where the magnetic field is changed from pointing in the x-direction and brought back to the same point after a rotation in the

$x - y$ plane. The lower energy wave function acquires a nontrivial Pancharatnam–Berry phase of $-\pi$.

While all this is okay, what has this to do with band insulators? The essential point is that R parameter we were discussing would also be the Bloch momentum index in a Brillouin zone! Therefore as we calculate such a phase over the momentum band, we can get a nontrivial phase. Such a momentum band, which hosts nontrivial phases are called topological bands. In a one-dimensional problem of SSH model where the momentum runs from $-\pi$ to π the Berry phase takes a value of π when $t_2 > |t_1|$ else zero. Therefore one can define an integer index C of the form $C = \gamma/\pi$ which takes a nontrivial value when $t_2 > |t_1|$ else is zero. This was essentially shown in Fig. 1.3. Similarly, an integral over a two dimensional surface is called the Chern number. Soon it was realized that the transverse conductivity σ_{xy} is in fact one such Berry phase [11],

$$\sigma_{xy} = \frac{e^2}{\hbar} \int \frac{d\mathbf{k}}{(2\pi)^2} F(\mathbf{k}) = n\frac{e^2}{h} \tag{1.23}$$

where n is an integer. This leads to quantized conductivity in integer quantum Hall effect. While in quantum Hall, magnetic field explicitly breaks time reversal symmetry, one can also have systems where no explicit magnetic field is present however one still has quantized Hall response since time reversal symmetry is broken due to peculiar band dispersion. Such systems are called Anomalous quantum Hall models ("anomalous" because no magnetic field is explicitly present.) Such a model was proposed by Haldane in a honeycomb lattice with (pseudo) spin-orbit coupling [15]. The Hamiltonian he devised [15] reads

$$H(\mathbf{k}) = 2t_2 \cos(\phi) \left(\sum_i \cos(\mathbf{k}.\mathbf{b}_i) \right) 1 + t_1 \left(\sum_i [\cos(\mathbf{k}.\mathbf{a}_i)\sigma_x + \sin(\mathbf{k}.\mathbf{a}_i)\sigma_y] \right)$$
$$+ \left[M - 2t_2 \sin\phi \left(\sum_i \sin(\mathbf{k}.\mathbf{b}_i) \right) \right] \sigma_z \tag{1.24}$$

here t_1 and t_2 are the nearest and next-to-nearest neighbor hoppings and ϕ is the phase in the next-to-nearest neighbor hopping. $\mathbf{a}, \mathbf{b}$ are lattice vectors as shown in Fig. 1.9. σs are the Pauli spin matrices, here they represent the two-orbital unit cell basis of the honeycomb lattice. The essential point to note is the corresponding phase diagram, as shown in Fig. 1.9, where Chern number takes nontrivial value in parameter regime of M and ϕ.

So what is interesting if the bands are topological, have nonzero winding number or Chern numbers? Here comes the essential idea of bulk-edge correspondence. To have a well defined bulk, the lattice has to be placed under periodic boundary conditions. Under these conditions the nontrivial topological numbers can be calculated. However, once such a lattice is placed under open boundary conditions, as is naturally true for a physical system, the boundary will host robust edge states. These edge states are special as they are *chiral* i.e. they move in only one direction, and

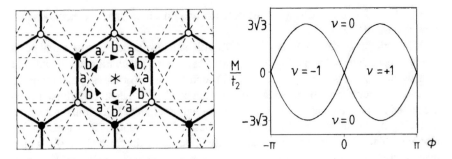

Fig. 1.9 The Haldane model: The lattice and the phase diagram which was proposed by Haldane in his 1988 paper. Adapted from [15]

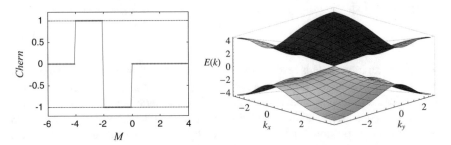

Fig. 1.10 Chern number: (Left) Chern number for the BHZ model as a function of M (see Eq. (1.25)). B is set to 1. (Right) The band dispersion for the BHZ model at $M = 0$. The linearly dispersing bands are clearly visible at $(k_x, k_y) = (0, 0)$

therefore cannot be scattered. Any conductance due to such edge states will be robust. Absence of any states which lie at the same spatial region which can host current in the opposite direction, imply that an impurity cannot scatter the original states. This is the origin of the topological protection. Such edge states are also called anomalous theories because, they live in a $d - 1$ dimensional space, but their effective Hamiltonians cannot be independently realized in any physically meaningful way just in $d - 1$ dimensions.

In favor of concreteness we restate the ideas discussed above in the case of a two dimensional Chern-insulator model (called the BHZ model after Bernevig–Hughes–Zhang) which will also be useful in one of the later chapters. Consider the following model for a Chern insulator on a square lattice (see [40–42] and references there in)

$$H = \sigma_x \sin k_x + \sigma_y \sin k_y + B(M + 2 - \cos k_x - \cos k_y)\sigma_z. \tag{1.25}$$

Here B and M are two parameters. One can set $B = 1$ and this model is known to have topological transitions in different regimes of M when the filling in the system is kept fixed at half (i.e., the lower band is filled). The regimes are,

$M > 0$	Trivial
$-2 < M < 0$	Topological I
$-4 < M < -2$	Topological II
$M < -4$	Trivial

Note that the spectrum of this model is symmetric about $E = 0$. In presence of periodic boundary conditions one can calculate the Chern number for the lower band of this model and the result is shown in Fig. 1.10. This is calculated using the numerical prescription of [43]. As one can notice that the transitions in Chern number coincide with the phases mentioned in the table above.

It is particularly interesting to expand the Hamiltonian at $M = 0$ and look at small k. The Hamiltonian looks like

$$H_{M \to 0} \approx \sigma_x k_x + \sigma_y k_y + M \sigma_z. \tag{1.26}$$

This is reminiscent of the spin problem in magnetic field we had discussed. This realization of a Dirac spectrum is the key to various topological phases in many systems and in fact therefore one expects as M switches from positive to negative, such Berry phase transitions will lead to Chern number change as is indeed also seen in Fig. 1.10.

In fact for this square lattice model, each of these topological transitions are due to realization of Dirac cones at different regions of the Brillouin zone as a function of M. The essential band closings as a function of M are shown in Fig. 1.11. The system is gapped at any M except at few values of M. At $M = 0$ there is a Dirac band closing

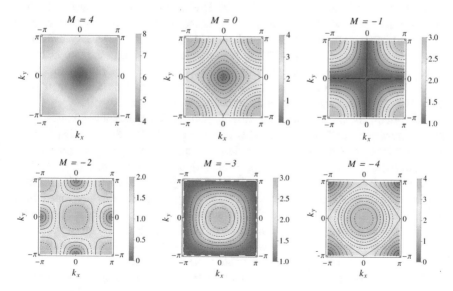

Fig. 1.11 The gap closings: The color shows the band dispersion of the positive energy band as a function of M for the BHZ model (see Eq. (1.25))

at $(k_x, k_y) = (0, 0)$. At $M = -2$ there are again band closings at $(k_x, k_y) = (0, \pm\pi)$ and $(k_x, k_y) = (\pm\pi, 0)$. At $M = -4$ there are band closings at $(k_x, k_y) = (\pm\pi, \pm\pi)$. Note that at these same values of M topological phase transitions occur (see Fig. 1.10). In the regimes where the system is topological, if one now imposes open boundary conditions, one expects existence of edge states due to bulk-edge correspondence. This is shown in Fig. 1.12. One can notice that one obtains midgap states in presence of open boundary conditions which live on the edge of the sample.

There is an alternate prescription to think about topological systems. Any system can be considered to be topologically trivial when its constituent sites are infinitely far i.e., when the sites *do not* couple. This means that all the wave functions are site-localized and at any filling corresponding number of Wannier orbitals will be localized. This is a trivial insulator. Therefore, any system which can be adiabatically connected to this trivial insulator is topologically trivial. While, systems which *cannot* be smoothly brought to this trivial insulator should be considered as a topological phase. In fact it is impossible to construct a Wannier like localized orbitals from a topological band without gap closing transitions [44, 45]. These ideas can be made more precise numerically and leads to a quantized topological index, which we will discuss in one of the later chapters.

We can notice in Eq. (1.24) and in Eq. (1.25) that the Hamiltonians were multi-component objects (due to presence of Pauli spin matrices). In the case of BHZ model the two-component wave function at every site can be considered as s and p orbital while it is the sublattice label in case of Haldane model. In both Haldane model and the BHZ model, the time reversal symmetry is broken and nowhere do we worry about the actual spin. However, extensions of both the BHZ model and the Haldane model can be made to the spinful case [16, 17, 46]. Here each spin does an "integer-quantum Hall" of its own kind and therefore the system in total can host quantum-*spin* Hall effect. Here again the edge states remain protected. In presence of time reversal symmetry, on the same edge there are two states which move in opposite directions, but now due to spin-momentum locking, the two electrons have two distinct spin-orientation. This implies that a usual impurity cannot scatter an incoming electron on the same edge. These class of materials, and their three dimensional generalizations are called topological insulators [47–49].

Such two-component objects can also be realized in superconductors where the "two componentness" is due to the particle and hole sector. Slowly what became clear is that there are a variety of models/systems where topological phenomena can manifest. Also such models can exist in various spatial dimensions. One of the puzzles which then appeared—"is there a relationship between such topological phenomena?" In 2008–09 a classification of all topological classes was developed. First in three dimensions [21] and then Kitaev showed a particular periodicity amongst topological phenomena and spatial dimension [20]. In this classification non-ordinary symmetries took the center stage which we will discuss in quite some detail in the next chapter. But for completion, we mention the basic results.

There exists three non-ordinary symmetries: time reversal, charge conjugation and sublattice, based on which *all* quadratic fermionic Hamiltonians can be classified. This is called the tenfold way [19, 50, 51]. Topological phases, on the other hand,

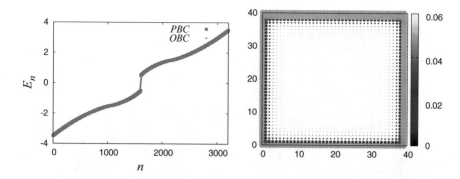

Fig. 1.12 The edge: (Left) The eigenenergies for a 40×40 system is shown with periodic boundary conditions (PBC) and with open boundary conditions (OBC) ($M = -0.5$) for the model described in (1.25). Midgap states appear when OBC is applied. (Right) The wave function corresponding to the central eigenvalue is plotted on the right. The blob size and color represents the probability of finding the particle there. The state is clearly edge localized

Table 1.1 The periodic table of topological insulators and superconductors. Adapted from [20, 50]

Class	(T,C,S)	$d = 0$	$d = 1$	$d = 2$	$d = 3$	$d = 4$	$d = 5$	$d = 6$	$d = 7$
A	(0,0,0)	$\mathbb{Z}$	0	$\mathbb{Z}$	0	$\mathbb{Z}$	0	$\mathbb{Z}$	0
AIII	(0,0,1)	0	$\mathbb{Z}$	0	$\mathbb{Z}$	0	$\mathbb{Z}$	0	$\mathbb{Z}$
AI	(+1,0,0)	$\mathbb{Z}$	0	0	0	$2\mathbb{Z}$	0	$\mathbb{Z}_2$	$\mathbb{Z}_2$
BDI	(+1,+1,1)	$\mathbb{Z}_2$	$\mathbb{Z}$	0	0	0	$2\mathbb{Z}$	0	$\mathbb{Z}_2$
D	(0,+1,0)	$\mathbb{Z}_2$	$\mathbb{Z}_2$	$\mathbb{Z}$	0	0	0	$2\mathbb{Z}$	0
DIII	(−1,+1,1)	0	$\mathbb{Z}_2$	$\mathbb{Z}_2$	$\mathbb{Z}$	0	0	0	$2\mathbb{Z}$
AII	(−1,0,0)	$2\mathbb{Z}$	0	$\mathbb{Z}_2$	$\mathbb{Z}_2$	$\mathbb{Z}$	0	0	0
CII	(−1,−1,1)	0	$2\mathbb{Z}$	0	$\mathbb{Z}_2$	$\mathbb{Z}_2$	$\mathbb{Z}$	0	0
C	(0,−1,0)	0	0	$2\mathbb{Z}$	0	$\mathbb{Z}_2$	$\mathbb{Z}_2$	$\mathbb{Z}$	0
CI	(+1,−1,1)	0	0	0	$2\mathbb{Z}$	0	$\mathbb{Z}_2$	$\mathbb{Z}_2$	$\mathbb{Z}$

can be characterized by the nature of the topological invariants. While in some systems, topological invariant such as Chern number can take all integer values, such a topological number will be classified as $\mathbb{Z}$. While in some other cases it can take only two values thereby meaning $\mathbb{Z}_2$. While in some others it can take all even integers—$2\mathbb{Z}$. Based on these a periodic table of topological phases in tenfold symmetry classes was presented [20, 21, 52]. This is shown in Table 1.1. Topological model based systems now fall in various classes of this table. For example, quantum Hall is in class A in $d = 2$, three-dimensional topological insulators are in class AII in $d = 3$.

While in this limited scope of introduction an attempt has been made to delve into ideas and topics which will be useful in the upcoming material in this thesis, the subject by itself is far reaching and has many underpinnings. In fact, there is a

Fig. 1.13 The pre-1987 condensed matter physics scenario: (Left) A colleague presents to Anderson a sketch (pre-1984) where Anderson is shown to be pondering about various problems including Si:P, Landau level physics, and many others. The fractional quantum Hall skyline can be seen on the horizon. (Right) Baskaran writes to Anderson, in 1987, regarding RVB theory and its uses and importance to variety of fields including High-Tc superconductors, heavy fermions etc. Adapted from [56]

fast developing thought paradigm to classify various phases on the basis of a concept called "topological order" [53]—an idea which we will hardly visit in this thesis. Briefly, in topologically ordered phases it is the long range entanglement that gives rise to various exotic properties of the system—such as robust ground state degeneracies when the system is placed on various manifolds, fractionalized excitations etc. Topological phases which we will discuss in this thesis are "topological", yet *do not* have topological order. These subtle ideas have been reviewed in a recent article [53]. For a more elaborate discussion on the tenfold way and related topological phases, interested reader is suggested the review articles [50, 51] and references therein. There are also many experimental breakthroughs which we donot discuss here, however can be found in [54, 55] and references therein.

1.5   The Landscape: *Ill* Condensed Matter

As was mentioned before, the phrase—*Ill* condensed matter, was first used as early as 1978 in a school which discussed disorder physics, percolation, glass etc. Why is the title of the thesis "Excursions in *ill* condensed matter"? To understand this, lets ask a question: What are the main problems condensed matter theorists have been worrying about over the years?

Its interesting to study a (pre-1984) drawing (see Fig. 1.13), which a colleague of Anderson presented to him. One can notice what used to occupy condensed matter theorists then. Anderson is seen wondering "What shall I plant next?" in his garden which has local moments, Si:P; He seems to sit on the disorder landscape and a wave function localized on it. The quantum Hall trace of conductance with magnetic field has appeared on the horizon. Interestingly in 1987, Baskaran sends a letter to Anderson, asking "Are we bees?" Baskaran is referring to the RVB theory which they developed for High-T_c superconductors [57] and how it has found applications in variety of strongly correlated systems, Heavy fermions etc.

Strong correlations, disorder, impurities seem to be the order of the day.

What is the situation today?

Not very different—treating interactions, disorder and impurities in a single framework is still extremely difficult. While major advances have been made, through multiple approaches—be it analytical or computational, we are still pondering and confused about some questions which were posed more than fifty years earlier. But, the major advancements in the field—topological physics, the realization of nonordinary symmetries etc.—have now enriched our platter and brought to us many more pertinent questions which need to be addressed.

The topological physics that we discussed in the last section, is built on translationally invariant systems—clean lattice models. Just as we had seen for few examples, the concept of bands and Brillouin zone, is in fact necessary to define a topological number in most situations.

What happens to a topological phase on an amorphous system?

Can we have a topological "glass"?

We also visited the Kitaev's table and saw the classification of topological phases based on various dimensions.

But—can we construct a topological model on a system which does not even have an integer dimension?

Earlier, community has worried about percolation—a phenomena where removal of lattice sites/bonds can inhibit conductivity—one of the topics which Thouless discussed in the 1978 school. We saw that scaling theory predicts that in two dimensions, *any* infinitesimal disorder inhibits conductivity in two dimensions.

What happens to a topological phase, as one shoots off the lattice sites or bonds?

What happens to the Hall conductance? Does it immediately become zero?

As the scope of condensed matter has expanded; the scope of *ill* condensed matter has also expanded. And in this thesis, we define it, to comprise of all the developments which have enriched the field—since then, till now. If anything, they are bound by shared abhorrence for *clean systems*.

Clean systems, are rendered unclean, not just by *ill* formed lattices, or *ill* formed bonds—sometimes outside agents can make it unclean. This part of *ill* condensed matter is the *impurity physics*. The surprise of the phenomena where few magnetic impurities can make a metal show insulating behavior falls in this category. Kondo effect, as we saw in the last section is the paradigmatic example which encompasses the essential physics of strong correlations, magnetism, asymptotic freedom etc.

Topological physics, as we know, mostly arises due to some crucial ingredients in the system. For example, in the case of quantum Hall it is the magnetic field. In many others, mostly lattice based systems, it is the spin-orbit coupling. In fact, in the last decade or so, spin-orbit coupling has emerged as an indispensable ingredient in realizing many exotic phenomena [58].

What happens to the Kondo effect, in presence of spin-orbit coupling?

Again a question, whose answer is not readily available.

The topological physics which we have been introduced to, is again understood mostly for noninteracting phases, where strong correlations do not play a role.

What happens to topological phases, in presence of interactions?

Some of the questions, mentioned above, we will address in this thesis. Some we will answer; for others we will take the crucial steps. Condensed matter field is a fast moving field, and the journey is set to become even more exciting. Since 1980s many answers to the questions which Anderson and Baskaran are worried about are still unanswered. But the developments in the last few decades have redrawn the edges. These new developments with the existing knowledge have now opened up the borders in condensed matter physics. And understanding systems away from the idealization of translational symmetry is now essential even for building the basic skeletal structures. Anderson's ideas of localization in disordered systems, have now questioned our fundamental understanding of thermalization [59]. Randomly coupled interacting motifs have shown non-Fermi liquid physics, which might have implications in various problems including in holography [60] . Kondo effect or broadly impurity physics—the progenitor of strong correlations—has transformed the first principles community, to develop and use more sophisticated and accurate tools to analyze strongly correlated materials [61]. Topology has also pervaded there for finding topological bands in realistic crystalline materials [62]. The ideas of topological phases, have brought the ideas of topological order into the field, where long-range entanglement can lead to fractionalized excitations, anyonic charge etc. [53]. Early discovery of fractional quantum Hall is now considered an example of such physics. Today, long range entanglement, and its spread is in fact one of the most actively pursued problems. Add to this story—gauge theories, dualities, spin liquids, frustrated magnetism. The sheer extent, variety and spread of the condensed matter arena is immense.

A beginner in the condensed matter field is not every different than an early explorer in undiscovered lands. The base camp is just a small oasis of clean systems—of idealized notions—where our Brillouin zones, Fermi surfaces etc. can rescue us. Some milestones and landmarks exist, but the real adventure lies in walking deeper into the wilderness of *ill* condensed matter. And this thesis presents some of these *excursions*. A vague landscape of ideas in condensed matter physics, clean or *ill*, is presented in Fig. 1.14.

Fig. 1.14   Clean and ill:
Schematic diagram of topics
in quantum condensed matter
and where they fit between
ill and clean condensed
matter physics. In blue are
some of the ideas which will
be discussed in this thesis

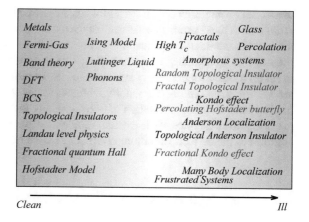

Metals   *Glass*

Fractals

Fermi-Gas   *Ising Model*   *High* T_c *Percolation*

Band theory   *Luttinger Liquid*   *Amorphous systems*

DFT   *Phonons*   *Random Topological Insulator*

Fractal Topological Insulator

BCS   *Kondo effect*

Percolating Hofstader butterfly

Topological Insulators

Anderson Localization

Landau level physics   *Topological Anderson Insulator*

Fractional quantum Hall   *Fractional Kondo effect*

Hofstadter Model   *Many Body Localization*

Frustrated Systems

Clean   *Ill*

1.6   The *Excursions*: Outline of This Thesis and Results

In the following, we briefly mention the works presented in this thesis. This thesis
is divided into eight chapters where the last one serves as the epilogue. The forthcoming six chapters discusses various problems and its solutions, all taken in spirit
of investigating physics of systems which are not translationally invariant. While
some sections are motivated on theoretical curiosities, some have deep implications
in material applications. Some of the chapters also provide mathematical formalism
and its applicability in physical settings. Many of these open up ample opportunities
for future work which we will briefly discuss in the last chapter.

1.6.1   Tenfold Way

As was mentioned before, it is known that based on three non-ordinary symmetries
one can classify all fermionic Hamiltonians into ten symmetry classes. Presently,
these symmetries are imposed on a case-by-case basis and it is not straight forward
to construct a generic Hamiltonian in any symmetry class. This has handicapped
the condensed matter community to be limited to specific models, mostly based on
lattices, and has rendered it quite challenging to set up symmetry-respecting Hamiltonians on arbitrary settings such as—polymers, molecules, or other systems which
can as well be *ill*. In this chapter we aim to remove this handicap. We provide a systematic treatment of the tenfold way of classifying noninteracting fermionic systems,
demanding no structure from the basic degrees of freedom. We identify four types of
symmetries that such systems can possess—one, ordinary type (usual unitary symmetries), and three *non*-ordinary symmetries (i.e., time reversal, charge conjugation
and sublattice). Focusing on systems that possess no nontrivial ordinary symmetries,
we demonstrate that the non-ordinary symmetries are strongly constrained, and using

these one can naturally obtain the tenfold classes. We also derive the canonical representations of these symmetries in each of the ten classes which enables us to construct the generic noninteracting quadratic Hamiltonians in each class. This provides the prescription to take any arbitrary setting—be it a random system, a polymer, and to set up a Hamiltonian in any symmetry class! In order to reproduce results which are known in literature—we rederive the conditions these symmetry classes apply on the momentum representation of translationally invariant Hamiltonians. We also provide some examples of the existent topological Hamiltonians and show how they conform to the derived Hamiltonian structures we have obtained.

1.6.2 Topological Insulators in Amorphous Systems

In the last chapter, we would have seen how to set up symmetry-respecting Hamiltonians on arbitrary motifs. Taking cues from this, we now set up *topological* Hamiltonians on randomly distributed sites in space, i.e., a random lattice and ask—Are topological phases possible in an amorphous system? In this chapter we establish such phases. To unequivocally prove the existence of this topological phase, we demonstrate robust edge states which can have quantized transport. We evaluate the Bott index, a topological number, and show that it is nontrivial in the topological phase. By tuning parameters such as the density of sites we achieve transitions from a trivial to a topological phase and propose a phase diagram. Additionally we investigate interesting features in the nature of topological phase transitions by system size scaling. To establish the generality of these results, we demonstrate this physics in all five classes (among the ten) in two dimensions, which are known to host topological phases. We also analyze a three dimensional system and demonstrate surface states in an amorphous setting. Our study not only provides a deeper understanding of the topological phases of noninteracting fermions, but also suggests some radically new directions in the pursuit of the laboratory realization of topological quantum matter. Till now, the community has been looking at crystalline systems to realize topological phases, but this chapter explores the possibility if glasses can be made topological. It opens up the possibility that appropriately doped semiconductors can turn topological.

1.6.3 Seeking Topological Phases in Fractals

In the last chapter, we would have established topological phases on amorphous systems. However, in these systems the dimension was still well defined, i.e., the sites existed either on a two dimensional plane or on a three dimensional box. Kitaev's classification of topological phases is also based on the non-ordinary symmetries and the spatial dimension. Can we have topological phases on systems, where the dimension is *not* an integer? First, do such systems even exist? The answer is *yes*.

Fractals are systems which can have non-integer dimensions. We construct topological Hamiltonians (i.e., Hamiltonians which realize topologically insulating phases on regular lattices) on fractals, where every site is equivalently coordinated (has same number of neighbors) as any other site. Due to which it is apriori unclear to demarcate a "bulk" and a "edge". We find that it is impossible to obtain a topologically "insulating" phase in such a system. These systems have a metallic phase in regime of parameters where usual lattice systems have topologically insulating phases. Typical wave functions close to band centre prefer some sites which we call as the "edges" of this system. We find that in appropriately defined thermodynamic limit, the fractal has bounded, self-similar, fractal "edge" spectrum. We connect this system to leads and study its two terminal transport. Interestingly these edges can conduct close to unit quantum of conductance! Moreover, these metals are chiral in nature. By this we mean that an excitation on any edge has a preferred direction to move in. These features point to a topological character in the system, yet, the Bott index is zero. We also investigate other fractals and higher dimensional generalizations. Our findings refuse to conform to the standard interpretation in terms of gapped topological phases and Kitaev's table. We report such phases in fractals and our results presses the question of how such phases should be classified. It will also be interesting to visit the ideas of bulk-boundary correspondence in context of these systems.

1.6.4 Killing the Hofstadter Butterfly

In the last chapter we would have seen a fractal spectrum arising from a topological system set up on a fractal. However, one does not necessarily need fractals for producing fractal spectrums. Electronic bands in a square lattice when subjected to a perpendicular magnetic field form the Hofstadter butterfly—a fractal pattern. This is a lattice analog of the quantum Hall effect and is in fact a topological model. One of the unanswered questions in the literature has been—how does removal of bonds affect this pattern? And what happens to the topological number? This is the question we address in this chapter.

  We study the evolution of this pattern as a function of bond percolation disorder (i.e., removal or dilution of lattice bonds). With increasing concentration of the bonds removed, the pattern gets smoothly decimated. However, in this process of decimation, bands develop interesting characteristics and features. For example, in the high disorder limit, some butterfly-like pattern still persists even as most of the states are localized. We also analyze, in the low disorder limit, the effect of percolation on wave functions (using inverse participation ratios) and on band gaps in the spectrum. We explain and provide the reasons behind many of the key features in our results by analyzing small clusters and finite size rings. Furthermore, we study the effect of bond dilution on transverse conductivity (σ_{xy}), which in this case is the "topological number". We show that starting from the clean limit, increasing disorder reduces σ_{xy} to zero, even though the strength of percolation is smaller than the classical percolation threshold. This shows that the system undergoes a direct transition from an integer quantum Hall state to a localized Anderson insulator beyond a critical value of bond dilution. We further find that the energy bands close to the band edge are more stable to disorder than at the band center. To arrive at these results we use the coupling matrix approach to calculate Chern numbers for disordered systems.

1.6.5 Fractional Spins and Kondo Effect

In the last couple of chapters we have focused on non-ordinary symmetries and their use in constructing topological phases in glassy and fractal systems. We also studied a model where a topological phase was annihilated by percolation. We now analyze another kind of *ill* condensed matter—systems with an impurity. Here one is interested in understanding how a correlated impurity affects the physics of an otherwise clean system.

In the last few decades, the important role of spin-orbit coupling has been increasingly appreciated. Systems with spin-orbit coupling have shown potential to realize exotic quantum states, be it topological phases or novel phenomena on the interfaces of strongly correlated systems. However, the role of an quantum impurity in such a system is still not well understood. We investigate the new physics that arises when a correlated spin-*1/2* quantum impurity hybridizes with a spin-orbit coupled Fermi system. We find that, in contrast to unit local moment in conventional Kondo effect, here, the impurity develops a *fractional local moment* of *2/3*. This implies that if one does an experiment (such as measuring susceptibility), it will seem that the spin of the electron is only $\frac{2}{3}$rd of its true value!

This local moment also results in a novel Kondo effect with a high Kondo temperature (T_K). We provide a theory that explains these novel features including the origins of the fractional local moment and provides a recipe to use spin-orbit coupling (λ) to enhance Kondo temperature ($T_K \sim \lambda^{4/3}$). We employ multiple techniques such as mean field theory, variational calculation and quantum Monte Carlo, to support and establish our results. Even as this finding of such unexpected and rich phenomena, in a simple looking many body system, is of interest in itself; It also points out opportunities for systems with tunable spin-orbit coupling (such as cold atoms) to explore these predictions experimentally.

1.6.6 Structure of Many-Body Hamiltonians in Different Symmetry Classes

In the earlier chapters we would have analyzed topological phases in noninteracting *ill* condensed matter systems. We would have also looked at impurity physics and a novel Kondo effect. One can ask, what happens to an arbitrarily interacting topological Hamiltonian on a random lattice?

While this is a honorable question, it is quite difficult to analyze such a problem. A complete classification of all topological phases of interacting fermions, broadly, is still a long standing open question.

In this chapter we provide possibly first few steps to even tackle such a problem. While the role of non-ordinary symmetries is now clear for noninteracting Hamiltonians (as described in Chap. 2), it is not yet clear, what will be a generic interacting Hamiltonian in each of these ten classes. This is the question we address in this chapter. We use the canonical representations of the symmetries and obtain the structure of the Hamiltonians with arbitrary N-body interactions in each of the ten classes. We show in detail the methodology of construction for arbitrary two-body Hamiltonians. We extend these results for generic N and see an interesting classification depending on whether the value of N is even or odd. We show that the space of N-body Hamiltonians which has either or both charge conjugation and sublattice symmetries puts interesting constraints on the Hamiltonian terms which have lower than N order interactions. We show an interesting result of many body *zero* modes in an interacting Hamiltonian which satisfies the sublattice symmetry. The results of this analysis is crucial through multiple directions. The results can can help address open questions on the topological classification of interacting fermionic systems. Our results can also be used to generate generic interacting Hamiltonians in ten classes and explore many-body localization physics.

These are the excursions described in this thesis. The concluding chapter provides a quick overview of the results and comments on future scope. Without further ado, in the next chapter we plunge into the discussion of ordinary and non-ordinary symmetries, culminating in the tenfold classification and some concrete examples.

References

1. Anderson PW (1972) More is different. Science 177(4047):393–396. arXiv:1011.1669v3
2. Marx K, Engels F (1987) Karl Marx and Frederick Engels collected works, vol 25. International Publishers
3. Wilhelm M, Mathison D, Cameron J (2009) Avatar: a confidential report on the biological and social history of Pandora. HarperCollins, UK
4. Abrahams E, Anderson PW, Licciardello DC, Ramakrishnan TV (1979) Scaling theory of localization: absence of quantum diffusion in two dimensions. Phys Rev Lett 42:673–676
5. Kondo J (1964) Resistance minimum in dilute magnetic alloys. Prog Theor Phys 32(1):37–49
6. Balian R, Maynard R, Toulouse G (1983) Ill-condensed matter, vol 31. World Scientific
7. Wilson KG (1975) The renormalization group: critical phenomena and the kondo problem. Rev Mod Phys 47:773–840
8. Klitzing KV, Dorda G, Pepper M (1980) New method for high-accuracy determination of the fine-structure constant based on quantized hall resistance. Phys Rev Lett 45:494–497
9. Von Klitzing K (1986) The quantized hall effect. Rev Mod Phys 58(3):519
10. Abbott BP, Abbott R et al (2016) Observation of gravitational waves from a binary black hole merger. Phys Rev Lett 116:061102
11. Thouless DJ, Kohmoto M, Nightingale MP, den Nijs M (1982) Quantized hall conductance in a two-dimensional periodic potential. Phys Rev Lett 49:405–408
12. Néel L (1971) Magnetism and local molecular field. Science 174(4013):985–992

13. Landau L (1930) Diamagnetismus der metalle. Zeitschrift für Physik 64(9–10):629–637
14. Su WP, Schrieffer JR, Heeger AJ (1980) Soliton excitations in polyacetylene. Phys Rev B 22:2099–2111
15. Haldane FDM (1988) Model for a quantum hall effect without Landau levels: Condensed-matter realization of the "parity anomaly". Phys Rev Lett 61:2015–2018
16. Kane CL, Mele EJ (2005) Z_2 topological order and the quantum spin hall effect. Phys Rev Lett 95:146802
17. Kane CL, Mele EJ (2005) Quantum spin hall effect in graphene. Phys Rev Lett 95:226801
18. Kitaev AY (2001) Unpaired majorana fermions in quantum wires. Physics-Uspekhi 44(10S):131
19. Altland A, Zirnbauer MR (1997) Nonstandard symmetry classes in mesoscopic normal-superconducting hybrid structures. Phys Rev B 55:1142–1161
20. Kitaev A (2009) Periodic table for topological insulators and superconductors. AIP Conf Proc 1134:22–30
21. Schnyder AP, Ryu S, Furusaki A, Ludwig AWW (2008) Classification of topological insulators and superconductors in three spatial dimensions. Phys Rev B 78:195125
22. Anderson PW (1958) Absence of diffusion in certain random lattices. Phys Rev 109:1492–1505
23. Mott NF, Davis EA (1971) Electronic processes in non-crystalline materials
24. Thouless DJ (1983) Percolation and localization. In: Balian R et al (eds) III-condensed matter: les houches session XXXI. Published by World Scientific Publishing Co. Pte. Ltd., pp 1–62. ISBN 9789814412728
25. Lee PA, Ramakrishnan TV (1985) Disordered electronic systems. Rev Mod Phys 57:287–337
26. Kramer B, MacKinnon A (1993) Localization: theory and experiment. Rep Prog Phys 56(12):1469
27. Ramakrishnan T (1987) Electron localization. In: Chance and matter, proceedings of the Les Houches summer school, session XLVI, pp 213–303
28. Markoš P (2006) Numerical analysis of the Anderson localization. Acta Phys Slovaca 56:561–685
29. Bulka B, Schreiber M, Kramer B (1987) Localization, quantum interference, and the metal-insulator transition. Zeitschrift fur Physik B Condensed Matter 66(1):21–30
30. Anderson PW (1961) Localized magnetic states in metals. Phys Rev 124:41–53
31. Hewson AC (1997) The Kondo problem to heavy fermions, vol 2. Cambridge University Press, Cambridge
32. Sarachik MP, Corenzwit E, Longinotti LD (1964) Resistivity of Mo-Nb and Mo-Re alloys containing 1% Fe. Phys Rev 135:A1041–A1045
33. Yosida K (1966) Bound state due to the s-d exchange interaction. Phys Rev 147:223–227
34. Anderson P (1970) A poor man's derivation of scaling laws for the kondo problem. J Phys C: Solid State Phys 3(12):2436
35. Krishna-murthy HR, Wilkins JW, Wilson KG (1980) Renormalization-group approach to the anderson model of dilute magnetic alloys. i. static properties for the symmetric case. Phys Rev B 21:1003–1043
36. Krishna-murthy HR, Wilkins JW, Wilson KG (1980) Renormalization-group approach to the anderson model of dilute magnetic alloys. ii. static properties for the asymmetric case. Phys Rev B 21:1044–1083
37. Hirsch JE, Fye RM (1986) Monte carlo method for magnetic impurities in metals. Phys Rev Lett 56:2521–2524
38. Sakurai JJ, Tuan S-F, Cummins ED (1995) Modern quantum mechanics, revised edn
39. Shankar R (2012) Principles of quantum mechanics. Springer Science & Business Media
40. Bernevig BA, Hughes TL (2013) Topological insulators and topological superconductors. Princeton University Press, Princeton
41. Shen S-Q (2013) Topological insulators: Dirac equation in condensed matters, vol 174. Springer Science & Business Media
42. Fradkin E (2013) Field theories of condensed matter physics. Cambridge University Press, Cambridge

43. Fukui T, Hatsugai Y, Suzuki H (2005) Chern numbers in discretized brillouin zone: efficient method of computing (spin) hall conductances. J Phys Soc Jpn 74(6):1674–1677
44. Thouless DJ (1984) Wannier functions for magnetic sub-bands. J Phys C: Solid State Phys 17(12):L325
45. Thonhauser T, Vanderbilt D (2006) Insulator/chern-insulator transition in the haldane model. Phys Rev B 74:235111
46. Bernevig BA, Hughes TL, Zhang S-C (2006) Quantum spin hall effect and topological phase transition in HgTe quantum wells. Science 314(5806):1757–1761
47. Hasan MZ, Kane CL (2010) Colloquium: topological insulators. Rev Mod Phys 82:3045–3067
48. Qi X-L, Zhang S-C (2011) Topological insulators and superconductors. Rev Mod Phys 83:1057–1110
49. Qi X-L, Zhang S-C (2010) The quantum spin Hall effect and topological insulators. Phys Today 63:33
50. Ludwig AWW (2016) Topological phases: classification of topological insulators and superconductors of non-interacting fermions, and beyond. Phys Scr 2016(T168):014001. arXiv:1512.08882
51. Chiu CK, Teo JCY, Schnyder AP, Ryu S (2016) Classification of topological quantum matter with symmetries. Rev Mod Phys 88:035005
52. Ryu S, Schnyder AP, Furusaki A, Ludwig AWW (2010) Topological insulators and superconductors: tenfold way and dimensional hierarchy. New J Phys 12(6):065010
53. Wen X-G (2016) Zoo of quantum-topological phases of matter, pp 1–16. arXiv:1610.03911
54. Hasan MZ, Kane CL (2010) Colloquium. Rev Mod Phys 82:3045–3067
55. Ando Y (2013) Topological insulator materials. J Phys Soc Jpn 82(10):102001
56. Baskaran G (2016) arXiv:1608.08587. In: Chandra P, Coleman P, Kotliar G, Ong P, Stein DL, Clare Yu (eds) PWA90 A life time of emergence (World Scientific 2016) and Anderson PW, in Modern physics in America—a Michaelson–Morley centennial symposium, Fickinger W, Kuwalski KL (eds) AIP conference proceedings 169 (American Institute of Physics, 1988)
57. Anderson PW, Baskaran G, Zou Z, Hsu T (1987) Resonating valence-bond theory of phase transitions and superconductivity in La_2Cuo_4-based compounds. Phys Rev Lett 58:2790–2793
58. Manchon A, Koo HC, Nitta J, Frolov SM, Duine RA (2015) New perspectives for rashba spin-orbit coupling. Nat Mater 14:871–882 (review)
59. Nandkishore R, Huse DA (2015) Many body localization and thermalization in quantum statistical mechanics. Ann Rev Condens Matter Phys 6(1):15–38
60. Kitaev A (2015) A simple model of quantum holography. Talks at KITP
61. Georges A, Kotliar G, Krauth W, Rozenberg MJ (1996) Dynamical mean-field theory of strongly correlated fermion systems and the limit of infinite dimensions. Rev Mod Phys 68:13–125
62. Bansil A, Lin H, Das T (2016) Colloquium. Rev Mod Phys 88:021004

Chapter 2
Tenfold Way

2.1 Introduction

In Chap. 1, we briefly discussed how symmetries and phases of matter are intertwined. In fact under Landau-Ginzburg-Wilson paradigm, classification of various phases are understood in terms of broken symmetries. In the last couple of decades, however, another set of symmetries have started playing most important role—time reversal, charge conjugation and sublattice symmetry. The advent of topological physics and its classification, as was discussed in Sect. 1.4.3, is deeply rooted in these symmetries. The essential moral of the story is that *every* fermionic Hamiltonian can be classified into a tenfold way through these three "intrinsic" symmetries [1–3]. This was a development over the seminal work of Dyson who had earlier classified Hamiltonians into a threefold way [4].

What are these symmetries? And their physical content? Condensed matter physicists are usually introduced to time reversal symmetry, when the idea of Kramer's degeneracies in half-integer spin systems is introduced. Through a subtle set of arguments (see [5]) one realizes that time reversal is in fact a *equation of motion* reversal operator and has to be antilinear. Physically motivated examples include cases where one expects that time reversal operator ($\mathscr{T}$) should convert

$$\mathscr{T} \, p \, \mathscr{T}^{-1} = -p \tag{2.1}$$

$$\mathscr{T} \, r \, \mathscr{T}^{-1} = r \tag{2.2}$$

where p and r are the momentum and position operators respectively. Similarly particle-hole is understood as a symmetry operation which takes a state in the conduction band to a state in the valence band or vice-versa. Sublattice symmetry is considered when the system can be differentiated into two parts—say A and B, and the Hamiltonian has only inter-A and B couplings and no intra-A and intra-B couplings. A square lattice or a honeycomb lattice with nearest neighbor hoppings are examples of such systems while a triangular lattice is not. A rule of thumb is that such a Hamiltonian, which satisfies the sublattice symmetry, should be writable in

© Springer Nature Switzerland AG 2019
A. Agarwala, *Excursions in Ill-Condensed Quantum Matter*,
Springer Theses, https://doi.org/10.1007/978-3-030-21511-8_2

the form

$$\mathbf{H} = \begin{pmatrix} \mathbf{0} & \mathbf{h}_{ab} \\ \mathbf{h}_{ab}^{\dagger} & \mathbf{0} \end{pmatrix}. \tag{2.3}$$

Why are there only three intrinsic symmetries? Moreover, these three symmetries seem qualitatively distinct, how is it that they can exhaust the complete classification? Why is this a "ten" fold way, and not more? How does one set up Hamiltonians in generic settings, say on a polymer? How does one then think about these symmetries physically?

   This chapter answers all these questions. We begin the discussion with the setting up of the stage in next few sections. This will be done *ab initio*. We first define what do we mean by a symmetry, and which operations are allowed. We then derive the four types of symmetry operations that a fermionic system can possess. One of these is the usual symmetry operations and the other three are non-ordinary symmetries. Section 2.6 introduces and studies those fermionic systems that possess no nontrivial ordinary symmetries and obtains the constraints that the symmetries have to satisfy. We identify that these three non-ordinary symmetries are in fact the time reversal, charge conjugation and sublattice symmetries. The tenfold way is elucidated in Sect. 2.7 which includes a treatment of the canonical representation of the symmetry in each of the symmetry classes, the result of which is displayed in Table 2.1. Next we will use these canonical forms to devise generic structures of Hamiltonians in each symmetry class. In this process we will rediscover the understanding of these symmetries. These generalized results are applied to translationally invariant systems and the conditions on momentum space representation of the Hamiltonians is derived. This will reproduce the known results in literature. Finally we end with some explicit examples where the role of these symmetries are manifest in canonical forms.

2.2   Framework

Consider a generic noninteracting fermionic Hamiltonian $\mathscr{H}$ written amongst L one particle states, $|i\rangle \equiv \psi_i^{\dagger}|0\rangle$, $i = 1, \ldots, L$ as,

$$\mathscr{H} = \sum_{ij} H_{ij} \psi_i^{\dagger} \psi_j \tag{2.4}$$

These one particle states are orthonormal $\langle i|j\rangle = \delta_{ij}$ (δ_{ij} is the Kronecker delta symbol) and can be created starting from the vacuum state $|0\rangle$ by use of creation operators ($\psi_i^{\dagger}$). Annihilation operators ($\equiv \psi_i$) destroys the ith state. ψ and $\psi^{\dagger}$s satisfy the fermion anticommutation relations

$$\psi_i \psi_j^{\dagger} + \psi_j^{\dagger} \psi_i = \{\psi_i, \psi_j^{\dagger}\} = \delta_{ij}, \tag{2.5}$$

and

$$\{\psi_i, \psi_j\} = 0. \tag{2.6}$$

Note that these states could denote a variety of situations; i could be orbitals at different sites of a lattice, or different atomic orbitals, or even flavor states of an elementary particle.

We can collect these fermionic operators into convenient arrays as,

$$\Psi = \begin{bmatrix} \psi_1 \\ \vdots \\ \psi_L \end{bmatrix}, \quad \Psi^\dagger = \begin{bmatrix} \psi_1^\dagger & \cdots & \psi_L^\dagger \end{bmatrix}. \tag{2.7}$$

and rewrite the Hamiltonian (see Eq. (2.4)) as,

$$\mathscr{H} = \Psi^\dagger \mathbf{H} \Psi = \begin{bmatrix} \psi_1^\dagger & \cdots & \psi_L^\dagger \end{bmatrix} \begin{bmatrix} H_{11} & \cdots & \cdots & H_{1L} \\ \vdots & \ddots & & \vdots \\ \vdots & & \ddots & \vdots \\ H_{L1} & \cdots & \cdots & H_{LL} \end{bmatrix} \begin{bmatrix} \psi_1 \\ \vdots \\ \psi_L \end{bmatrix} \tag{2.8}$$

where $\mathbf{H}$ represents the matrix form of the Hamiltonian which satisfies $\mathbf{H} = \mathbf{H}^\dagger$. One aim of this chapter is to understand how various symmetries constraint the structure of the matrix $\mathbf{H}$.

2.3 Usual Symmetry

So, when do we say that a system has a symmetry?

A symmetry operation as defined by Wigner [6] is a linear *or* antilinear operator $\mathscr{U}$ acting on the Hilbert space of the system that leaves the magnitude of the inner product invariant. Stated in the context of one particle states of our L-orbital fermionic system, an operator $\mathscr{U}$ is a symmetry operation if

$$|\langle \zeta' | \varphi' \rangle| = |\langle \zeta | \varphi \rangle|, \quad \forall \ |\varphi\rangle, |\zeta\rangle \in \mathcal{V}_1 \tag{2.9}$$

with $|\varphi'\rangle = \mathscr{U} |\varphi\rangle$ and $|\zeta'\rangle = \mathscr{U} |\zeta\rangle$ where $\mathcal{V}_1$ is the vector space of all one particle states.

Linear or antilinear denotes whether the symmetry commutes or anticommutes with $i \equiv \sqrt{-1}$.

$$\mathscr{U}(i\mathscr{I})\mathscr{U}^{-1} = i\mathscr{I} \quad linear \tag{2.10}$$

$$\mathscr{U}(i\mathscr{I})\mathscr{U}^{-1} = -i\mathscr{I} \quad antilinear \tag{2.11}$$

where $\mathscr{I}$ is the identity operator on $\mathcal{V}_1$. In terms of operators these symmetry operations keeps annihilation operators as annihilation operators, and creation operators as creation operators upto a unitary transformation, that is,

$$\mathscr{U}\Psi^\dagger\mathscr{U}^{-1} = \Psi^\dagger\mathbf{U} \tag{2.12}$$

$$\mathscr{U}\Psi\mathscr{U}^{-1} = \mathbf{U}^\dagger\Psi. \tag{2.13}$$

Under a linear symmetry transformation $\mathscr{H}$, as we had defined previously, transforms to

$$\mathscr{H} = \Psi^\dagger\mathbf{H}\Psi \rightarrow \Psi^\dagger\mathbf{U}\mathbf{H}\mathbf{U}^\dagger\Psi \quad linear \tag{2.14}$$

$$\mathscr{H} = \Psi^\dagger\mathbf{H}\Psi \rightarrow \Psi^\dagger\mathbf{U}\mathbf{H}^*\mathbf{U}^\dagger\Psi. \quad antilinear \tag{2.15}$$

A Hamiltonian is therefore said to have a symmetry if,

$$\Psi^\dagger\mathbf{U}\mathbf{H}\mathbf{U}^\dagger\Psi = \Psi^\dagger\mathbf{H}\Psi \quad linear \tag{2.16}$$

$$\Psi^\dagger\mathbf{U}\mathbf{H}^*\mathbf{U}^\dagger\Psi = \Psi^\dagger\mathbf{H}\Psi \quad antilinear \tag{2.17}$$

which provides the conditions

$$\mathbf{U}\mathbf{H}\mathbf{U}^\dagger = \mathbf{H} \quad or \quad [\mathbf{U}, \mathbf{H}] = 0 \quad linear \tag{2.18}$$

$$\mathbf{U}\mathbf{H}^*\mathbf{U}^\dagger = \mathbf{H} \quad or \quad [\mathbf{U}\mathbf{K}, \mathbf{H}] = 0 \quad antilinear \tag{2.19}$$

where $\mathbf{K}$ is the complex conjugation operator.

2.4 Transposing Symmetry

In order to introduce this idea, we need to first go beyond single particle states. We realize that our system can have any number of particles N_P ranging from 0 to L. For each particle number N_P, the set of allowed states are $\binom{L}{N_P}$ in number which can be obtained by creating N_P-particles from the vacuum using distinct combinations of the operators $\psi_i^\dagger$. Let us denote the vector space of N_P particle states as $\mathcal{V}_{N_P}$.

The full Hilbert-Fock space is therefore given by

$$\mathcal{V} = \bigoplus_{N_P=0}^{L} \mathcal{V}_{N_P}, \tag{2.20}$$

which is a vector space over complex numbers $\mathbb{C}$. What is interesting to note is that the number of states for $N_P = L - 1$ particles is same as that for one particle. This suggests that we can map every particle state to a corresponding "hole" state. In fact this mapping is valid for every N_P particle state to a $L - N_P$ particle state. Therefore

Fig. 2.1 Illustrating "usual" and "transposing" symmetry operations: A usual symmetry operation maps $\mathcal{V}_{N_P} \mapsto \mathcal{V}_{N_P}$, while a transposing symmetry operation maps $\mathcal{V}_{N_P} \mapsto \mathcal{V}_{L-N_P}$

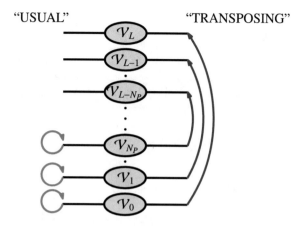

"USUAL"   "TRANSPOSING"

while a usual symmetry, of the kind described in the previous section, keeps one in the same particle sector, a "transposing" symmetry will take us from the N_P sector to $L - N_P$ sector. The above discussion can be summarized by the equations

$$\mathcal{U}_{\text{USL}}(\mathcal{V}_{N_P}) = \mathcal{V}_{N_P}, \ \forall N_P \tag{2.21}$$

$$\mathcal{U}_{\text{TRN}}(\mathcal{V}_{N_P}) = \mathcal{V}_{L-N_P}, \ \forall N_P. \tag{2.22}$$

Therefore, as Altland and Zirnbauer [1] had realized, many particle fermionic systems can in fact admit a larger classes of symmetry operations. A linear or antilinear operation that maps $\mathcal{V}_{N_P}$ to $\mathcal{V}_{L-N_P}$ preserving the magnitude of inner product and also the graded nature of the Hilbert-Fock space (see Eq. (2.20)) is also a legitimate symmetry operation. We call such a operation as *transposing* (see Fig. 2.1 for a schematic illustration).

We can implement the action of a transposing symmetry operation $\mathcal{U}_{\text{TRN}}$ via

$$\mathcal{U}_{\text{TRN}}\Psi^{\dagger}\mathcal{U}_{\text{TRN}}^{-1} = \Psi^{T}\mathbf{U}_{\text{TRN}}^{*} \tag{2.23}$$

where * denotes complex conjugation. Note that here, creation operators are mapped to annihilation operators and vice-versa. Also transposing symmetry operation can be linear or antilinear.

One can ask, when will a Hamiltonian satisfy a transposing symmetry? For such symmetries, the Hamiltonian $\mathcal{H}$ satisfies

$$: \mathcal{U}_{\text{TRN}}\mathcal{H}\mathcal{U}_{\text{TRN}}^{-1} := \mathcal{H} \tag{2.24}$$

where $:$ $:$ indicates that the expression $\mathcal{U}_{\text{TRN}}\mathcal{H}\mathcal{U}_{\text{TRN}}^{-1}$ has to be normal ordered (all creation operators to the left of annihilation operators). This procedure needs to be done since, as we had just seen, transposing symmetry operations converts annihilation operators to creation operators and vice versa. The conditions equivalent

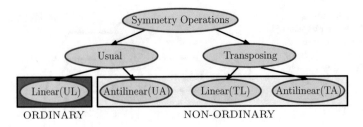

Fig. 2.2 Four types of symmetry operations

to Eqs. (2.16) and (2.17) becomes,

$$: \Psi^T \mathbf{U}^* \mathbf{H} (\mathbf{U}^*)^\dagger (\Psi^\dagger)^T : \ = \Psi^\dagger \mathbf{H} \Psi \quad linear \tag{2.25}$$

$$: \Psi^T \mathbf{U}^* \mathbf{H}^* (\mathbf{U}^*)^\dagger (\Psi^\dagger)^T : \ = \Psi^\dagger \mathbf{H} \Psi. \quad antilinear \tag{2.26}$$

This as we will see later will produce interesting conditions on **H**.

The main conclusion of the above discussion is that there are four distinct types of symmetry operations as illustrated in Fig. 2.2. They are usual linear (UL), usual antilinear (UA), transposing linear (TL), and transposing antilinear (TA). We find it useful to introduce additional terminology—we call UL symmetry operations as *ordinary* symmetry operations, and all other types of symmetry operations (UA, TL and TA) as *non-ordinary* operations.

2.5 Symmetry Group

The set of all symmetry operations $\mathscr{U}$ of the system forms a group $G_\mathcal{V}$. This group is a disjoint union of four types of symmetry operations (see Fig. 2.2),

$$G_\mathcal{V} = G_\mathcal{V}^{\mathrm{UL}} \cup G_\mathcal{V}^{\mathrm{UA}} \cup G_\mathcal{V}^{\mathrm{TL}} \cup G_\mathcal{V}^{\mathrm{TA}}. \tag{2.27}$$

The non-ordinary symmetries have the following interesting properties. First, for any non-ordinary $\mathscr{U}$ we have

$$\mathscr{U} \mathscr{U} \in G_\mathcal{V}^{\mathrm{UL}}, \quad \forall \mathscr{U} \in G_\mathcal{V}^{\mathrm{UA}} \cup G_\mathcal{V}^{\mathrm{TL}} \cup G_\mathcal{V}^{\mathrm{TA}}. \tag{2.28}$$

which can be paraphrased "the square of a non-ordinary symmetry operation is an ordinary symmetry operation". Second, the product of two distinct types of non-ordinary symmetry operations is the third type—for example

Fig. 2.3 Multiplication table of types of symmetries

$G_1 \downarrow \mid G_2 \rightarrow$	$G_{\mathcal{V}}^{\mathrm{UL}}$	$G_{\mathcal{V}}^{\mathrm{UA}}$	$G_{\mathcal{V}}^{\mathrm{TL}}$	$G_{\mathcal{V}}^{\mathrm{TA}}$
$G_{\mathcal{V}}^{\mathrm{UL}}$	$G_{\mathcal{V}}^{\mathrm{UL}}$	$G_{\mathcal{V}}^{\mathrm{UA}}$	$G_{\mathcal{V}}^{\mathrm{TL}}$	$G_{\mathcal{V}}^{\mathrm{TA}}$
$G_{\mathcal{V}}^{\mathrm{UA}}$	$G_{\mathcal{V}}^{\mathrm{UA}}$	$G_{\mathcal{V}}^{\mathrm{UL}}$	$G_{\mathcal{V}}^{\mathrm{TA}}$	$G_{\mathcal{V}}^{\mathrm{TL}}$
$G_{\mathcal{V}}^{\mathrm{TL}}$	$G_{\mathcal{V}}^{\mathrm{TL}}$	$G_{\mathcal{V}}^{\mathrm{TA}}$	$G_{\mathcal{V}}^{\mathrm{UL}}$	$G_{\mathcal{V}}^{\mathrm{UA}}$
$G_{\mathcal{V}}^{\mathrm{TA}}$	$G_{\mathcal{V}}^{\mathrm{TA}}$	$G_{\mathcal{V}}^{\mathrm{TL}}$	$G_{\mathcal{V}}^{\mathrm{UA}}$	$G_{\mathcal{V}}^{\mathrm{UL}}$

$$\mathscr{U}_{\mathrm{UA}} \mathscr{U}_{\mathrm{TL}} = \mathscr{U}_{\mathrm{TA}}; \tag{2.29}$$

this is summarized in Fig. 2.3.

2.6 Grotesque Fermionic Systems

We will now focus attention on a special class of L-orbital fermionic systems. Our systems of interest do not possess any nontrivial ordinary symmetries—we dub such systems as "*grotesque* fermionic systems (GFS)" to highlight this property. This would suggest that the only ordinary symmetry operation allowed is the trivial identity operation $\mathscr{I}$. However, since we are working with the Hilbert-Fock vector space (not a projective or "ray" space of Wigner, see [7]), the operator

$$\mathscr{I}_\theta = e^{\mathrm{i}\theta \mathscr{N}} \tag{2.30}$$

with

$$\mathscr{N} = \sum_{i=1}^{L} \psi_i^\dagger \psi_i \tag{2.31}$$

is always an allowed symmetry operation, and indeed *any* number conserving Hamiltonian will be invariant under this operation.

We will now demonstrate an important property of a GFS. A GFS can possess *at most* one each of UA, TL and TA symmetries. In other words, *non-ordinary symmetries of a GFS are solitary*. To prepare to prove this statement, we adopt some useful notation. We will denote UA symmetry operations by $\mathscr{T}$, TL by $\mathscr{C}$, and TA by $\mathscr{S}$— this choice anticipates later discussion. Suppose, now, that we have two distinct UA symmetries of the GFS, say $\mathscr{T}_1$ and $\mathscr{T}_2$, then $\mathscr{T}_1 \mathscr{T}_2$ is also a symmetry of our system. But from Fig. 2.3, we know that $\mathscr{T}_1 \mathscr{T}_2$ is an UL type symmetry. By definition, in a GFS any unitary symmetry has to be a trivial one, $\mathscr{I}_\theta$ for some θ. Thus $\mathscr{T}_1 \mathscr{T}_2 = \mathscr{I}_\theta$ or $\mathscr{T}_2 = \mathscr{I}_{-\theta} \mathscr{T}_1^{-1}$ and hence $\mathscr{T}_2$ is not a distinct UA symmetry—it is simply a product of a trivial symmetry with $\mathscr{T}_1^{-1}$. The same argument works for $\mathscr{C}$ and $\mathscr{S}$ symmetries (see also [8]).

Consider the symmetry operator $\mathscr{T}^2$. From the previous paragraph we know $\mathscr{T}^2 = \mathscr{I}_\theta$ for some θ. We can determine θ from $\mathscr{T}^{-1}\mathscr{T}^2 = \mathscr{T} = \mathscr{T}^2\mathscr{T}^{-1}$ which implies $\mathscr{I}_{-\theta}\mathscr{T}^{-1} = \mathscr{T} = \mathscr{I}_\theta\mathscr{T}^{-1}$. Applying this relation on $\mathcal{V}_1$ immediately forces $e^{i2\theta} = 1$ (due to the antilinearity of $\mathscr{T}$) or $e^{i\theta} = \pm 1$, and thus we immediately see that the action of $\mathscr{T}^2$ on $\mathcal{V}$ is

$$\mathscr{T}^2 = (\pm 1)^{\mathcal{N}}\,\mathscr{I}. \tag{2.32}$$

We will write Eq. (2.32) simply as

$$\mathscr{T}^2 = \pm\mathscr{I}. \tag{2.33}$$

We turn now to $\mathscr{C}^2$ which should also be equal to $\mathscr{I}_\theta$. We get, again, $\mathscr{C}^{-1}\mathscr{I}_\theta = \mathscr{I}_\theta\mathscr{C}^{-1}$. Applying this last relation on $\mathcal{V}_{N_P}$, we get

$$e^{iN_P\theta} = e^{i(L-N_P)\theta}, \quad \forall\, N_P = 0, \ldots, L. \tag{2.34}$$

resulting in $e^{i2\theta} = 1$, and

$$\mathscr{C}^2 = (\pm 1)^{\mathcal{N}}\,\mathscr{I}. \tag{2.35}$$

Note also that Eq. (2.34) implies that L must be even when $\theta = \pi$, i.e., a TL type of symmetry with the $(-)$ signature can only be implemented in a GFS with even number of orbitals.

Finally, we discuss $\mathscr{S}^2$. Noting that $\mathscr{S}$ is a TA symmetry, we get

$$e^{-iN_P\theta} = e^{i(L-N_P)\theta}, \quad \forall\, N_P = 0, \ldots, L\ , \tag{2.36}$$

and thus $e^{iL\theta} = 1$, implies $\theta = \frac{2\pi\ell}{L}$ where ℓ is one of $0, 1, \ldots, L-1$. We will later show that θ can always be chosen to be zero, leading to

$$\mathscr{S}^2 = \mathscr{I}. \tag{2.37}$$

We conclude this discussion on the properties of GFS symmetries by noting that if a GFS has a $\mathscr{T}$ and $\mathscr{C}$ type symmetry, then $\mathscr{S}$ is equal to $\mathscr{T}\mathscr{C}$. If a GFS has a $\mathscr{T}$ symmetry, we say that it possesses *time reversal symmetry*, $\mathscr{C}$ implies the presence of *charge conjugation symmetry*, and $\mathscr{S}$ endows a *sublattice symmetry* on the GFS.

2.7 The Tenfold Way

In this section, we show how the ten symmetry classes of fermions arise and obtain the canonical representations of the symmetries in these classes.

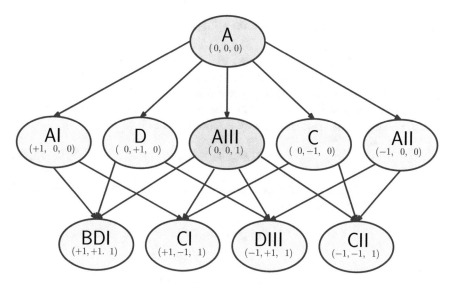

Fig. 2.4 Ten symmetry classes, and their interrelationships: The two classes shaded in red are
the complex classes, while all others are real classes. The symmetry signature is denoted by a triple
(T, C, S) as discussed in the text

2.7.1 Symmetry Classes

Based on the discussion of the previous section, we see that a GFS has to be of one
of three types.

Type 0 Possesses no non-ordinary symmetries.
Type 1 Possesses one non-ordinary symmetry.
Type 3 Possesses all three non-ordinary symmetries.

The resulting symmetry classes and the class hierarchy is shown in Fig. 2.4.

Whenever time reversal is present it be realized as $\mathscr{T}^2 = \pm\mathscr{I}$, and we denote
this by T $= \pm 1$, similarly $\mathscr{C}^2 = \pm\mathscr{I}$ is denoted by C $= \pm 1$ and presence of $\mathscr{S}$
is shown by S $= 1$. Absence of these symmetries is denoted by T $= 0$, C $= 0$, or
S $= 0$ as the case may be. The "symmetry signature" of any class is denoted by
a triple (T, C, S) (see Fig. 2.4). It is now immediately clear that there is only one
class of type 0 GFS, called A with symmetry signature $(0, 0, 0)$. There are five
classes of type 1 with a single non-ordinary symmetry: AI$(+1, 0, 0)$, AII$(-1, 0, 0)$,
D$(0, +1, 0)$, C$(0, -1, 0)$ and AIII$(0, 0, 1)$. The type 3 systems come in four classes:
BDI$(+1, +1, 1)$, CI$(+1, -1, 1)$, DIII$(-1, +1, 1)$, and CII$(-1, -1, 1)$. These
exhausts all the ten symmetry classes.

2.7.2   Canonical Representation of Symmetries

An important step towards determining the structure of Hamiltonians in each of these ten symmetry classes is the determination of the canonical representation of the symmetries of each class. This question is addressed in this section.

Each of the non-ordinary symmetry operations, $\mathscr{T}$, $\mathscr{C}$ and $\mathscr{S}$ is represented by a $L \times L$ unitary matrix. Time reversal $\mathscr{T}$ is represented as

$$\mathscr{T}\Psi^\dagger \mathscr{T}^{-1} = \Psi^\dagger \mathbf{U_T}. \qquad (2.38)$$

Charge conjugation and sublattice, being transposing operations, are represented using Eq. (2.23) respectively as

$$\mathscr{C}\Psi^\dagger \mathscr{C}^{-1} = \Psi^T \mathbf{U_C^*} \qquad (2.39)$$

and

$$\mathscr{S}\Psi^\dagger \mathscr{S}^{-1} = \Psi^T \mathbf{U_S^*}. \qquad (2.40)$$

2.7.2.1   Type 0

Class A: Class A has no non-ordinary symmetries and hence nothing to represent. There are, therefore, no restrictions on the L-orbital systems—any L orbital system can be in class A.

2.7.2.2   Type 1

Classes AI and AII: Time reversal symmetry is the sole non-ordinary symmetry present in these classes with $T = \pm 1$. The condition Eq. (2.33) gives (using the antilinearity of $\mathscr{T}$)

$$\mathscr{T}^2 \Psi^\dagger \mathscr{T}^{-2} = \Psi^\dagger \mathbf{U_T}\mathbf{U_T^*} = T\,\Psi^\dagger \qquad (2.41)$$

leading to

$$\mathbf{U_T}\mathbf{U_T^*} = T\,\mathbf{1} \qquad (2.42)$$

where $\mathbf{1}$ is an $L \times L$ unit matrix. An immediate consequence of this is that $|\det \mathbf{U_T}|^2 = T^L$, leading us to the conclusion that $T = +1$ can be realized in any L-orbital GFS, while $T = -1$ requires L to be an even number.

To construct a canonical $\mathbf{U}_T$ that satisfies Eq. (2.42), we consider a change of basis of the GFS (as $\widetilde{\Psi} = \mathbf{R}^{\dagger}\Psi$). The unitary $\widetilde{\mathbf{U}}_T$ representing $\mathscr{T}$ in this new basis can be obtained from

$$\mathbf{R}\widetilde{\mathbf{U}}_T\mathbf{R}^T = \mathbf{U}_T . \tag{2.43}$$

It is known that any unitary that satisfies Eq. (2.42) with $T = +1$ can be written as $\mathbf{U}_T = \mathbf{A}\mathbf{A}^T$ where $\mathbf{A}$ is a unitary matrix. This result from matrix theory, usually called the Takagi decomposition (see Appendix D of [9]), allows us to conclude that we can always choose $(\mathbf{R} = \mathbf{A})$ a basis[1] where $\widetilde{\mathbf{U}}_T = \mathbf{1}$ for $T = +1$. The outcome of this discussion is that $T = +1$ admits a canonical representation $\mathbf{U}_T = \mathbf{1}$.

In the case of $T = -1$, Takagi decomposition provides that any $\mathbf{U}_T$ satisfying Eq. (2.42) can decomposed as $\mathbf{A}\mathbf{J}\mathbf{A}^T = \mathbf{U}_T$ where

$$\mathbf{J} = \begin{pmatrix} \mathbf{0}_{MM} & \mathbf{1}_{MM} \\ -\mathbf{1}_{MM} & \mathbf{0}_{MM} \end{pmatrix}, \quad L = 2M, \quad \mathbf{J}\mathbf{J}^* = -\mathbf{1}, \tag{2.44}$$

and $\mathbf{A}$ is unitary. This is consistent with the fact that L must be necessarily even for $T = -1$ as concluded above. The subscripts on the matrices denote their sizes. Taken together with Eq. (2.43) allows us to conclude that $T = -1$ case is canonically represented by $\mathbf{U}_T = \mathbf{J}$.

Classes D and C: These classes respectively with symmetry signatures $(0, +1, 0)$ and $(0, -1, 0)$ possess the sole non-ordinary symmetry of charge conjugation. As noted just after Eq. (2.35) the latter symmetry can be realized only in GFSs with even number of orbitals.

As in the previous para, considering the action of $\mathscr{C}^2$ on the fermion operators, owing to Eq. (2.35), gives

$$\mathscr{C}^2\Psi^{\dagger}\mathscr{C}^{-2} = \mathscr{C}\Psi^T\mathbf{U}_C^*\mathscr{C}^{-1} = \Psi^{\dagger}\mathbf{U}_C\mathbf{U}_C^* \tag{2.45}$$

resulting in

$$\mathbf{U}_C\mathbf{U}_C^* = C\,\mathbf{1}. \tag{2.46}$$

Note that this relation, despite $\mathscr{C}$ being a *linear* operation, looks very similar to Eq. (2.42) of an usual *antilinear* operator.

Under a change of basis, it can be shown that $\mathbf{U}_C$ transforms in exactly the same manner as $\mathbf{U}_T$, i.e., via Eq. (2.43). Precisely the same considerations using Takagi decomposition of the previous para then allow us to conclude that there is a basis for the GFS where $\mathbf{U}_C$ can be represented canonically as $\mathbf{U}_C = \mathbf{1}$ for $C = +1$ and $\mathbf{U}_C = \mathbf{J}$ for $C = -1$.

---

[1]This new basis is *not* unique. In fact, existence of such a basis implies that any other basis related by a real orthogonal matrix will also be an equally valid one.

Considering a quadratic Hamiltonian provides that

$$- \mathbf{U_C} [\mathbf{H}]^T \mathbf{U_C^\dagger} = \mathbf{H} \tag{2.47}$$

which can be rewritten as

$$- \mathbf{CHC}^{-1} = \mathbf{H}. \tag{2.48}$$

This is consistent with the first quantized form

$$\mathbf{C} = \mathbf{U_C K}. \tag{2.49}$$

It is easily seen that this feature is generic—any transposing linear symmetry operation has an *antilinear* first quantized representation.

Class AIII: The sole non-ordinary symmetry in this class with the symmetry signature $(0, 0, 1)$ is the sublattice symmetry.

Investigating the action of $\mathscr{S}^2$, noting that $\mathscr{S}$ is an antilinear operator, we obtain from Eq. (2.36)

$$\mathscr{S}^2 \Psi^\dagger \mathscr{S}^{-2} = \mathscr{S} \Psi^T \mathbf{U_S^*} \mathscr{S}^{-1} = \Psi^\dagger \mathbf{U_S U_S} = e^{i \frac{2\pi\ell}{L}} \Psi^\dagger \tag{2.50}$$

implying

$$\mathbf{U_S U_S} = e^{i \frac{2\pi\ell}{L}} \mathbf{1}. \tag{2.51}$$

It is evident that it can be redefined ($e^{-i\frac{\pi\ell}{L}} \mathbf{U_S} \mapsto \mathbf{U_S}$) as

$$\mathbf{U_S U_S} = \mathbf{1}. \tag{2.52}$$

Since $\mathbf{U_S}$ is unitary, we obtain that $\mathbf{U_S^\dagger} = \mathbf{U_S}$, and thus, $\mathbf{U_S}$ is Hermitian. This condition implies that all eigenvalues of $\mathbf{U_S}$ are real and of unit magnitude. Quite interestingly, under a change of basis, $\mathbf{U_S}$ transforms as

$$\mathbf{R \tilde{U}_S R^\dagger} = \mathbf{U_S} \tag{2.53}$$

This implies that there is a basis in which $\mathbf{U_S}$ has the following canonical form

$$\mathbf{U_S} = \mathbf{1}_{p,q} \tag{2.54}$$

where

$$\mathbf{1}_{p,q} = \begin{pmatrix} \mathbf{1}_{pp} & \mathbf{0}_{pq} \\ \mathbf{0}_{qp} & -\mathbf{1}_{qq} \end{pmatrix} \tag{2.55}$$

where $p + q = L$. This development makes the meaning of the sublattice symmetry clear. The orbitals of the system are divided into two groups—"sublattices A and B". The sublattice symmetry operation, a transposing antilinear operation, maps the "particle states" in the A orbitals to "hole states" on A orbitals, while B-particle states are mapped to *negative B*-hole states. In the first quantized language, the transposing antilinear operator $\mathscr{S}$, is, therefore, represented by a *linear* matrix, i.e.,

$$S = U_S \tag{2.56}$$

and in a noninteracting system the symmetry is realized when

$$- SHS^{-1} = H \tag{2.57}$$

2.7.2.3 Type 3

The remainder of the four classes are of type 3, i.e., they possess all three non-ordinary symmetries. As noted at the end of Sect. 2.6, $\mathscr{S} = \mathscr{T}\mathscr{C}$ in these classes implying that

$$U_S = U_T U_C^* . \tag{2.58}$$

Our strategy in analyzing these classes is to choose a basis where $U_S = \mathbf{1}_{p,q}$, and to determine the structure of U_T and U_C. In this basis, we can write

$$U_T = \begin{pmatrix} \mathbf{u}_{pp} & \mathbf{u}_{pq} \\ T\mathbf{u}_{pq}^T & \mathbf{u}_{qq} \end{pmatrix}, \quad \mathbf{u}_{pp}^T = T\mathbf{u}_{pp}, \mathbf{u}_{qq}^T = T\mathbf{u}_{qq},$$

$$\text{for} \quad T = \pm 1, \tag{2.59}$$

where **u**s are complex matrices of the dimension indicated by the suffix. This form automatically satisfies Eq. (2.42). Once, we fix U_T, U_C is fixed for every class in type 3 via Eq. (2.58), as

$$U_C = U_T^T U_S^* = \begin{pmatrix} T\mathbf{u}_{pp} & -T\mathbf{u}_{pq} \\ \mathbf{u}_{pq}^T & -T\mathbf{u}_{qq} \end{pmatrix}; \quad T = \pm 1 . \tag{2.60}$$

The conditions Eqs. (2.42) and (2.46) constrain $\mathbf{u}_{pp}, \mathbf{u}_{pq}, \mathbf{u}_{qq}$ of Eq. (2.59) very strongly. Two possible cases arise.

T = C: This case results in the condition $U_T U_T^* = U_C U_C^*$ which provides the following constraints

$$\begin{aligned}
\mathbf{u}_{pp}\mathbf{u}_{pp}^* &= T\mathbf{1}_{pp}, & \mathbf{u}_{qq}\mathbf{u}_{qq}^* &= T\mathbf{1}_{qq}, \\
\mathbf{u}_{pq}\mathbf{u}_{pq}^\dagger &= \mathbf{0}_{pp}, & \mathbf{u}_{pq}^T\mathbf{u}_{pq}^* &= \mathbf{0}_{qq}, \\
\mathbf{u}_{pp}\mathbf{u}_{pq}^* &= \mathbf{0}_{pq}, & \mathbf{u}_{pq}^T\mathbf{u}_{pp}^* &= \mathbf{0}_{qp}, \\
\mathbf{u}_{pq}\mathbf{u}_{qq}^* &= \mathbf{0}_{pq}, & \mathbf{u}_{qq}\mathbf{u}_{pq}^\dagger &= \mathbf{0}_{qp}, & \text{for T} = \text{C}.
\end{aligned} \tag{2.61}$$

$T = -C$: Note that since either T or C is -1, L is already even $L = 2M$. The condition $\mathbf{U}_T\mathbf{U}_T^* = -\mathbf{U}_C\mathbf{U}_C^*$ obtains the constraints

$$
\begin{aligned}
\mathbf{u}_{pp}\mathbf{u}_{pp}^* &= \mathbf{0}_{pp}, & \mathbf{u}_{qq}\mathbf{u}_{qq}^* &= \mathbf{0}_{qq}, \\
\mathbf{u}_{pq}\mathbf{u}_{pq}^\dagger &= \mathbf{1}_{pp}, & \mathbf{u}_{pq}^T\mathbf{u}_{pq}^* &= \mathbf{1}_{qq}, \\
\mathbf{u}_{pp}\mathbf{u}_{pq}^* &= \mathbf{0}_{pq}, & \mathbf{u}_{pq}^T\mathbf{u}_{pp}^* &= \mathbf{0}_{qp}, \\
\mathbf{u}_{pq}\mathbf{u}_{qq}^* &= \mathbf{0}_{pq}, & \mathbf{u}_{qq}\mathbf{u}_{pq}^\dagger &= \mathbf{0}_{qp}, & \text{for } T = -C.
\end{aligned}
\tag{2.62}
$$

The second line of Eq. (2.62), forces $p = q = M$ and $\mathbf{u}_{pq}$ to be a $M \times M$ unitary matrix which we will call $\mathbf{u}_{MM}$. The other conditions provide $\mathbf{u}_{pp} = \mathbf{0}_{pp}$ and $\mathbf{u}_{qq} = \mathbf{0}_{qq}$.

Class BDI: The class has a symmetry signature $(1, 1, 1)$. Since $T = C = +1$, we see from Eq. (2.61) that $\mathbf{u}_{pq} = \mathbf{0}_{pq}$, along with $\mathbf{u}_{pp}\mathbf{u}_{pp}^* = +\mathbf{1}_{pp}$ and $\mathbf{u}_{qq}\mathbf{u}_{qq}^* = +\mathbf{1}_{qq}$, providing

$$
\mathbf{U}_T = \begin{pmatrix} \mathbf{u}_{pp} & \mathbf{0}_{pq} \\ \mathbf{0}_{qp} & \mathbf{u}_{qq} \end{pmatrix}, \quad \mathbf{U}_C = \begin{pmatrix} \mathbf{u}_{pp} & \mathbf{0}_{pq} \\ \mathbf{0}_{qp} & -\mathbf{u}_{qq} \end{pmatrix}
\tag{2.63}
$$

Apart from $p + q = L$, there are no additional constraints on p and q. The natural splitting of the orbitals into a p-subspace and q-subspace now allows us to find canonical forms of $\mathbf{U}_T$ and $\mathbf{U}_C$. Again, Takagi decomposition allows us to find a unitary matrix $\mathbf{r}_{pp}$ such that $\mathbf{r}_{pp}\mathbf{r}_{pp}^T = \mathbf{u}_{pp}$, another similar matrix $\mathbf{r}_{qq}$ that does $\mathbf{r}_{qq}\mathbf{r}_{qq}^T = \mathbf{u}_{qq}$. We can define a basis change matrix

$$
\mathbf{R} = \begin{pmatrix} \mathbf{r}_{pp} & \mathbf{0}_{pq} \\ \mathbf{0}_{qp} & \mathbf{r}_{qq} \end{pmatrix}
\tag{2.64}
$$

from which we immediately see that

$$
\mathbf{U}_T = \mathbf{R}\mathbf{1}\mathbf{R}^T, \quad \mathbf{U}_C = \mathbf{R}(\mathbf{1}_{p,q})\mathbf{R}^T.
\tag{2.65}
$$

Moreover, we see that $\mathbf{U}_S$ transforms under this basis change via Eq. (2.53) as

$$
\mathbf{R}^\dagger\mathbf{1}_{p,q}\mathbf{R} = \mathbf{1}_{p,q}
\tag{2.66}
$$

i.e., such a basis change does not alter the form of $\mathbf{U}_S$. The conclusion of this discussion is that when $T = C = +1$, we can always choose a basis in which

$$
\mathbf{U}_T = \mathbf{1}, \quad \mathbf{U}_C = \mathbf{1}_{p,q}, \quad \mathbf{U}_S = \mathbf{1}_{p,q}.
\tag{2.67}
$$

Class CII: This symmetry class has signature $(-1, -1, 1)$. From Eq. (2.61), we have $\mathbf{u}_{pq} = \mathbf{0}_{pq}$, $\mathbf{u}_{pp}\mathbf{u}_{pp}^* = -\mathbf{1}_{pp}$ and $\mathbf{u}_{qq}\mathbf{u}_{qq}^* = -\mathbf{1}_{qq}$, resulting in

$$U_T = \begin{pmatrix} \mathbf{u}_{pp} & \mathbf{0}_{pq} \\ \mathbf{0}_{qp} & \mathbf{u}_{qq} \end{pmatrix}, \quad U_C = \begin{pmatrix} -\mathbf{u}_{pp} & \mathbf{0}_{pq} \\ \mathbf{0}_{qp} & \mathbf{u}_{qq} \end{pmatrix}. \tag{2.68}$$

Even as $p + q = L = 2M$, there are additional constraints on p and q. The conditions $\mathbf{u}_{pp}\mathbf{u}_{pp}^* = -\mathbf{1}_{pp}$ and $\mathbf{u}_{qq}\mathbf{u}_{qq}^* = -\mathbf{1}_{qq}$, force $p = 2r$ and $q = 2s$, i.e., both p and q have to be even numbers, $2(r + s) = 2M$.

We can find canonical forms for U_T and U_C much like Eq. (2.67). For this, noting that p and q are even, we know that there are unitary matrices $\mathbf{r}_{pp}$ and $\mathbf{r}_{qq}$ such that $\mathbf{r}_{pp}\mathbf{J}_{pp}\mathbf{r}_{pp}^T = \mathbf{u}_{pp}$ and $\mathbf{r}_{qq}\mathbf{J}_{qq}\mathbf{r}_{qq}^T = \mathbf{u}_{qq}$. Thus one can perform a basis change using a transformation similar to Eq. (2.64) to obtain natural forms

$$U_T = \begin{pmatrix} \mathbf{J}_{pp} & \mathbf{0}_{pq} \\ \mathbf{0}_{qp} & \mathbf{J}_{qq} \end{pmatrix}, \quad U_C = \begin{pmatrix} -\mathbf{J}_{pp} & \mathbf{0}_{pq} \\ \mathbf{0}_{qp} & \mathbf{J}_{qq} \end{pmatrix}, \quad U_S = \mathbf{1}_{p,q}, \tag{2.69}$$

which provide the canonical representation of the three symmetries in class CII.

Class CI: This class with the symmetry signature $(+1, -1, 1)$ can be analyzed using Eq. (2.62) which provides

$$U_T = \begin{pmatrix} \mathbf{0}_{MM} & \mathbf{u}_{MM} \\ \mathbf{u}_{MM}^T & \mathbf{0}_{MM} \end{pmatrix}, \quad U_C = \begin{pmatrix} \mathbf{0}_{MM} & -\mathbf{u}_{MM} \\ \mathbf{u}_{MM}^T & \mathbf{0}_{MM} \end{pmatrix} \tag{2.70}$$

We can obtain canonical forms of U_T and U_C by the following manipulations. Consider a basis transformation of the type Eq. (2.64) of the from

$$\mathbf{R} = \begin{pmatrix} \mathbf{u}_{MM} & \mathbf{0}_{MM} \\ \mathbf{0}_{MM} & \mathbf{1}_{MM} \end{pmatrix} \tag{2.71}$$

This basis change effects the following changes (see Eqs. (2.43), (2.53))

$$U_T \mapsto \mathbf{R}^\dagger U_T \mathbf{R}^* = \begin{pmatrix} \mathbf{0}_{MM} & \mathbf{1}_{MM} \\ \mathbf{1}_{MM} & \mathbf{0}_{MM} \end{pmatrix} \equiv \mathbf{F}$$

$$U_C \mapsto \mathbf{R}^\dagger U_C \mathbf{R}^* = -\mathbf{J}$$

$$U_S \mapsto \mathbf{R}^\dagger \mathbf{1}_{M,M} \mathbf{R} = \mathbf{1}_{M,M} \tag{2.72}$$

where the first line defines the $L \times L$ matrix $\mathbf{F}$. Thus the canonical forms of class CII are

$$U_T = \mathbf{F}, \quad U_C = -\mathbf{J}, \quad U_S = \mathbf{1}_{M,M}. \tag{2.73}$$

Class DIII: The symmetry signature $(-1, +1, 1)$ provides $T = -1 = -C$. Form Eq. (2.62), we get

$$U_T = \begin{pmatrix} \mathbf{0}_{MM} & \mathbf{u}_{MM} \\ -\mathbf{u}_{MM}^T & \mathbf{0}_{MM} \end{pmatrix}, \quad U_C = \begin{pmatrix} \mathbf{0}_{MM} & \mathbf{u}_{MM} \\ \mathbf{u}_{MM}^T & \mathbf{0}_{MM} \end{pmatrix} \tag{2.74}$$

Table 2.1 Canonical representations of symmetry operators $\mathbf{U}_T$, $\mathbf{U}_C$ and $\mathbf{U}_S$ are shown in the ten symmetry classes. $\mathbf{1}$ is the $L \times L$ identity matrix, $\mathbf{J}$ defined in Eq. (2.44), $\mathbf{1}_{p,q}$ defined in Eq. (2.55), $\mathbf{F}$ defined in Eq. (2.72)

Class	T	C	S	L	$\mathbf{U}_T$	$\mathbf{U}_C$	$\mathbf{U}_S$
A	0	0	0	L	–	–	–
AI	+1	0	0	L	$\mathbf{1}$	–	–
AII	−1	0	0	$L = 2M$	$\mathbf{J}$	–	–
D	0	+1	0	L	–	$\mathbf{1}$	–
C	0	−1	0	$L = 2M$	–	$\mathbf{J}$	–
AIII	0	0	1	$L = p + q$	–	–	$\mathbf{1}_{p,q}$
BDI	+1	+1	1	$L = p + q$	$\mathbf{1}$	$\mathbf{1}_{p,q}$	$\mathbf{1}_{p,q}$
CII	−1	−1	1	$L = p + q$ $p = 2r;\ q = 2s$	$\begin{pmatrix} \mathbf{J}_{pp} & \mathbf{0}_{pq} \\ \mathbf{0}_{qp} & \mathbf{J}_{qq} \end{pmatrix}$	$\begin{pmatrix} -\mathbf{J}_{pp} & \mathbf{0}_{pq} \\ \mathbf{0}_{qp} & \mathbf{J}_{qq} \end{pmatrix}$	$\mathbf{1}_{p,q}$
CI	+1	−1	1	$L = 2M$	$\mathbf{F}$	$-\mathbf{J}$	$\mathbf{1}_{M,M}$
DIII	−1	+1	1	$L = 2M$	$\mathbf{J}$	$\mathbf{F}$	$\mathbf{1}_{M,M}$

Using the basis change Eq. (2.71), we can find canonical structure of the $\mathbf{U}$ matrices as

$$\mathbf{U}_T = \mathbf{J}, \quad \mathbf{U}_C = \mathbf{F}, \quad \mathbf{U}_S = \mathbf{1}_{M,M}. \tag{2.75}$$

In the classes CI and DIII, there is no further constraint on M which can be odd or even.

The determination of the canonical structure of the symmetry operation accomplishes a great deal in the determination of the structure of the Hamiltonians in each class. The summary of the findings of this section are given in Table 2.1.

2.8 Noninteracting Systems

We devote this section to obtaining the structure of the Hamiltonians of each class when there are no interactions present. Here the Hamiltonian can be considered to be composed of two matrices $\mathbf{H} = \left(\mathbf{H}^{(0)}, \mathbf{H}^{(1)} \right)$, the constant and the quadratic Hamiltonian respectively. Another important quantity of interest is Schödinger time evolution operator

$$\mathbb{U}_{\text{Schröd}}(t) = e^{-it\mathbb{H}} \tag{2.76}$$

where t is the time, $\mathbb{H}$ is the Hamiltonian matrix constructed out of $\mathbf{H} = \left(\mathbf{H}^{(0)}, \mathbf{H}^{(1)} \right)$. For example, the in this noninteracting setting the components of $\mathbb{H}_{ij} = H^{(0)} \delta_{ij} + H^{(1)}_{i;j}$. For a fixed time, say $t = 1$, the time evolution operator spans out a "geometric structure" as $\mathbf{H}$ runs over all of the space $\mathcal{H}$. As pointed out by [1] (generalizing the pioneering work of Dyson [4]), this geometric structure realized is a *symmetric space* (see [10] for a review symmetric spaces from a physicist's perspective, [11] for a review of Lie groups and algebras). The ten classes realize the the ten different symmetric spaces classified by Cartan, and in fact the names of the classes borrows Cartan's nomenclature. Our development of the ten fold symmetry classification not only allows us to recover these known results in a simple and direct manner but also provides a very clear structure of the Hamiltonian space that could be particularly useful for model building. Our results are recorded in Table 2.2.

Under the action of usual symmetries **H** transforms as

$$\mathbf{H} = \left(\mathbf{H}^{(0)}, \mathbf{H}^{(1)}\right) \mapsto \mathring{\mathbf{H}} = (\mathbf{H}^{(0)}, \check{\mathbf{H}}^{(1)}) \tag{2.77}$$

and for transposing symmetries we obtain

$$\mathbf{H} = \left(\mathbf{H}^{(0)}, \mathbf{H}^{(1)}\right) \mapsto \mathring{\mathbf{H}} = \left(\mathbf{H}^{(0)} + \mathrm{tr}_1 \check{\mathbf{H}}^{(1)}, -\left[\check{\mathbf{H}}^{(1)}\right]^T\right) \tag{2.78}$$

The structure of **H** is obtained by imposing the symmetry conditions for all the appropriate symmetries. An immediate consequence is that $\mathbf{H}^{(0)}$, the "vacuum energy", is allowed to have any real value $H^{(0)}$ in *all* the ten classes. This is a feature that is always true, i.e., even when interactions are present. As the values of H^0 spans the reals, the time evolution operator (at a fixed time) acquires a phase factor represented by the $U(1)$ group (see discussion on class A in the next para).

Class A: There are no constraints on the Hamiltonian in this class. Any $L \times L$ Hermitian matrix is in $\mathcal{H}_A^{(1)}$. Consequently $\mathrm{i}\mathcal{H}_A^{(1)} = \mathfrak{u}(L)$, and $\mathbb{U}_{\mathrm{Schröd}}(t)$ acquires the structure of $U(1) \times U(L)$. The $U(1)$ factor arises from $\mathbf{H}^{(0)}$ which spans all of real numbers. As discussed in the previous para, this $U(1)$ factor will appear in *all* cases (interacting and noninteracting). Henceforth, we suppress this factor (which is equivalent to setting $H^0 = 0$ (or any other fixed real number). The symmetric space of $\mathbb{U}_{\mathrm{Schröd}}(t)$ in Table 2.2 for this case, therefore is shown just as $U(L)$.

Class AI: For this class (with $\mathbf{U}_T = \mathbf{1}$) provides that real symmetric matrices $\mathbf{H}^{(1)} = [\mathbf{H}^{(1)}]^*$ are the allowed entries of $\mathcal{H}_{AI}^{(1)}$. This implies that the $\mathrm{i}\mathcal{H}_{AI}^{(1)} = \mathfrak{u}(L) \setminus \mathfrak{o}(L)$, where $\mathfrak{o}(L)$ is Lie algebra of the group of orthogonal matrices. The time evolution operator $\mathbb{U}_{\mathrm{Schröd}}(t)$ spans the coset space $U(L)/O(L)$.

Class AII: Since $\mathbf{U}_T = \mathbf{J}$ in this class with $L = 2M$, we can split the orbitals into two types and write the Hamiltonian as

$$\mathbf{H}^{(1)} = \begin{pmatrix} \mathbf{h}_{aa} & \mathbf{h}_{ab} \\ \mathbf{h}_{ab}^\dagger & \mathbf{h}_{bb} \end{pmatrix}. \tag{2.79}$$

The symmetry condition gives $\mathbf{J}[\mathbf{H}^{(1)}]^*\mathbf{J}^\dagger = \mathbf{H}^{(1)}$ resulting in $\mathbf{h}_{ab} = -\mathbf{h}_{ab}^T$, $\mathbf{h}_{bb} = \mathbf{h}_{aa}^*$. Here, $\mathbf{h}_{aa}, \mathbf{h}_{ab}, \mathbf{h}_{bb}$ are $M \times M$ matrices. The space $\mathcal{H}_{AII}$ is made of matrices of the type

$$\mathbf{H}^{(1)} = \begin{pmatrix} \mathbf{h}_{aa} & \mathbf{h}_{ab} \\ -\mathbf{h}_{ab}^* & \mathbf{h}_{aa}^* \end{pmatrix}. \tag{2.80}$$

It can be seen by explicit calculation that $i\mathcal{H}_{\text{AII}}^{(1)} = \mathbf{u}(2M) \setminus \mathbf{usp}(2M)$, leading to the time evolution $\mathbb{U}_{\text{Schröd}}(t)$ spanning the coset space $U(2M)/USp(2M)$. Here $USp(2M)$ is the symplectic group and $\mathbf{usp}(2M)$ is its associated Lie algebra.

Class D: This class with the charge conjugation described by $\mathbf{U}_C = \mathbf{1}$ is made of Hamiltonians with vanishing trace $\text{tr}\, \mathbf{H}^{(1)} = 0$ that satisfy, $\mathbf{H}^{(1)} = -[\mathbf{H}^{(1)}]^*$. The crucial distinction between the time reversal class $\text{AI}(T = +1)$ is that the transposing nature of the charge conjugation give the negative sign in the just stated symmetry condition. The space $i\mathcal{H}_{\text{D}}^{(1)}$ is made of real antisymmetric matrices which is the Lie algebra $\mathbf{o}(L)$. The time evolution operator spans $O(L)$.

Class C: The charge conjugation operation is described by $\mathbf{U}_C = \mathbf{J}$, and the symmetry condition $\mathbf{J}[\mathbf{H}^{(1)}]^* \mathbf{J}^\dagger = -\mathbf{H}^{(1)}$, leads to matrices with $\text{tr}\, \mathbf{H}^{(1)} = 0$,

$$\mathbf{H}^{(1)} = \begin{pmatrix} \mathbf{h}_{aa} & \mathbf{h}_{ab} \\ \mathbf{h}_{ab}^* & -\mathbf{h}_{aa}^* \end{pmatrix}, \tag{2.81}$$

i.e., $i\mathcal{H}_C^{(1)} = \mathbf{usp}(2M)$, and $\mathbb{U}_{\text{Schröd}}(t)$ spans $USp(2M)$.

Class AIII: The physics of this class is governed by the sublattice symmetry represented by the unitary $\mathbf{U}_S = \mathbf{1}_{p,q}$ that naturally partitions the orbitals into two groups (sublattices). It is natural to write the Hamiltonian as

$$\mathbf{H}^{(1)} = \begin{pmatrix} \mathbf{h}_{pp} & \mathbf{h}_{pq} \\ \mathbf{h}_{pq}^\dagger & \mathbf{h}_{qq} \end{pmatrix}. \tag{2.82}$$

The symmetry condition leads immediately to $\mathbf{h}_{pp} = \mathbf{0}_{pp}$, $\mathbf{h}_{qq} = \mathbf{0}_{qq}$. Thus, $\mathcal{H}_{\text{AIII}}$ is made of matrices of the form

$$\mathbf{H}^{(1)} = \begin{pmatrix} \mathbf{0}_{pp} & \mathbf{h}_{pq} \\ \mathbf{h}_{pq}^\dagger & \mathbf{0}_{qq} \end{pmatrix}. \tag{2.83}$$

Indeed, we have $i\mathcal{H}_{\text{AIII}}^{(1)} = \mathbf{u}(p+q) \setminus (\mathbf{u}(p) \oplus \mathbf{u}(q))$, resulting in $\mathbb{U}_{\text{Schröd}}(t)$ being in the coset space $U(p+q)/(U(p) \times U(q))$. It is interesting to note that this approach has reproduced the structure of Hamiltonian we had discussed in Eq. (2.3), and the corresponding intuitive understanding of sublattice symmetry is manifest. All of the type 3 classes can be viewed as descendants of class AIII, and thus all of these classes—often referred to as chiral classes—have Hamiltonians that imbibe the structure in Eq. (2.83).

Class BDI: The symmetry condition Eq. (2.77) with $\mathbf{U}_T = \mathbf{1}$ gives $\mathbf{h}_{pq} = \mathbf{h}_{pq}^*$, and thus elements of $\mathcal{H}_{\text{BDI}}$ are of the form

$$\mathbf{H}^{(1)} = \begin{pmatrix} \mathbf{0}_{pp} & \mathbf{h}_{pq} \\ \mathbf{h}_{pq}^T & \mathbf{0}_{qq} \end{pmatrix}, \quad \mathbf{h}_{pq} = \mathbf{h}_{pq}^* \tag{2.84}$$

We obtain $i\mathcal{H}_{\mathrm{BDI}}^{(1)} = \mathfrak{o}(p+q) \setminus (\mathfrak{o}(p) \oplus \mathfrak{o}(q))$, with $\mathbb{U}_{\mathrm{Schröd}}(t)$ spanning the the coset space $O(p+q)/(O(p) \times O(q))$.

Class CII: The symmetry condition Eq. (2.77) with $\mathbf{U}_T$ as shown in Table 2.1 and $\mathbf{H}^{(1)}$ of the form Eq. (2.83) provides

$$\mathbf{J}_{pp}\mathbf{h}_{pq}^* = \mathbf{h}_{pq}\mathbf{J}_{qq}. \tag{2.85}$$

The key point in this class is that the p orbitals themselves arise as $2r$ orbitals and q as $2s$ orbitals. Thus, $\mathbf{h}_{pq}$ can be written as

$$\mathbf{h}_{pq} = \begin{pmatrix} \mathbf{h}_{aa} & \mathbf{h}_{ab} \\ \mathbf{h}_{ba} & \mathbf{h}_{bb} \end{pmatrix}. \tag{2.86}$$

$\mathbf{h}_{aa}, \mathbf{h}_{ab}, \mathbf{h}_{ba}, \mathbf{h}_{bb}$ are $r \times s$ matrices. The condition Eq. (2.85) implies $\mathbf{h}_{ba} = -\mathbf{h}_{ab}^*$ and $\mathbf{h}_{bb} = \mathbf{h}_{aa}^*$ resulting in

$$\mathbf{h}_{pq} = \begin{pmatrix} \mathbf{h}_{aa} & \mathbf{h}_{ab} \\ -\mathbf{h}_{ab}^* & \mathbf{h}_{aa}^* \end{pmatrix}. \tag{2.87}$$

We see that $i\mathcal{H}_{\mathrm{CII}}^{(1)} = \mathfrak{usp}(2(r+s)) \setminus (\mathfrak{usp}(2r)) \oplus \mathfrak{usp}(2s))$ and the symmetric space generated by the time evolution operator is $USp(2(r+s))/(USp(2r) \times USp(2s))$.

Class CI: Here $L = 2M$, and thus Eq. (2.83) provides

$$\mathbf{H}^{(1)} = \begin{pmatrix} \mathbf{0}_{MM} & \mathbf{h}_{MM} \\ \mathbf{h}_{MM}^\dagger & \mathbf{0}_{MM} \end{pmatrix}. \tag{2.88}$$

The symmetry condition Eq. (2.77) with $\mathbf{U}_T = \mathbf{F}$ gives $\mathbf{h}_{MM}^T = \mathbf{h}_{MM}$ with

$$\mathbf{H}^{(1)} = \begin{pmatrix} \mathbf{0}_{MM} & \mathbf{h}_{MM} \\ \mathbf{h}_{MM}^* & \mathbf{0}_{MM} \end{pmatrix}. \tag{2.89}$$

It is now clear that the $i\mathcal{H}_{\mathrm{CI}}^{(1)} = \mathfrak{usp}(2M) \setminus \mathfrak{usp}(M)$, and the symmetric space corresponding to the time evolution is the coset space $USp(2M)/U(M)$.

Class DIII: In this $2M$ dimensional GFS with, the symmetry condition Eq. (2.77) leads to $\mathbf{h}_{MM}^T = -\mathbf{h}_{MM}$, and

$$\mathbf{H}^{(1)} = \begin{pmatrix} \mathbf{0}_{MM} & \mathbf{h}_{MM} \\ -\mathbf{h}_{MM}^* & \mathbf{0}_{MM} \end{pmatrix} \tag{2.90}$$

Here $i\mathcal{H}_{\mathrm{DIII}}^{(1)}$ is isomorphic to $\mathfrak{o}(2M) \setminus \mathfrak{u}(M)$. The symmetric space spanned by the time evolution operator is $O(2M)/U(M)$.

Table 2.2 Structure of noninteracting Hamiltonians in the ten symmetry classes

Class	L	$\mathbf{H}^{(1)}$	dim $i\mathcal{H}^{(1)}$	$i\mathcal{H}^{(1)}$	$\mathbb{U}_{\text{Schröd}}(t)$
A(0, 0, 0)	L	$\mathbf{H}^{(1)} = [\mathbf{H}^{(1)}]^{\dagger}$	L^2	$\mathbf{u}(L)$	$U(L)$
AI(+1, 0, 0)	L	$\mathbf{H}^{(1)} = [\mathbf{H}^{(1)}]^{*}$	$L(L+1)/2$	$\mathbf{u}(L) \setminus \mathbf{o}(L)$	$U(L)/O(L)$
AII(−1, 0, 0)	$L = 2M$	$\begin{pmatrix} \mathbf{h}_{aa} & \mathbf{h}_{ab} \\ -\mathbf{h}_{ab}^{*} & \mathbf{h}_{aa}^{*} \end{pmatrix}$	$M(2M-1)$	$\mathbf{u}(2M) \setminus \mathbf{usp}(2M)$	$U(2M)/USp(2M)$
D(0, +1, 0)	L	$\mathbf{H}^{(1)} = -[\mathbf{H}^{(1)}]^{*}$	$L(L-1)/2$	$\mathbf{o}(L)$	$O(L)$
C(0, −1, 0)	$L = 2M$	$\begin{pmatrix} \mathbf{h}_{aa} & \mathbf{h}_{ab} \\ \mathbf{h}_{ab}^{*} & -\mathbf{h}_{aa}^{*} \end{pmatrix}$	$M(2M+1)$	$\mathbf{usp}(2M)$	$USp(2M)$
AIII(0, 0, 1)	$L = p+q$	$\begin{pmatrix} \mathbf{0}_{pp} & \mathbf{h}_{pq} \\ \mathbf{h}_{pq}^{\dagger} & \mathbf{0}_{qq} \end{pmatrix}$	$2pq$	$\mathbf{u}(p+q) \setminus (\mathbf{u}(p) \oplus \mathbf{u}(q))$	$U(p+q)/(U(p) \times U(q))$
BDI(+1, +1, 1)	$L = p+q$	$\begin{pmatrix} \mathbf{0}_{pp} & \mathbf{h}_{pq} \\ \mathbf{h}_{pq}^{T} & \mathbf{0}_{qq} \end{pmatrix}$, $\mathbf{h}_{pq}^{*} = \mathbf{h}_{pq}$	pq	$\mathbf{o}(p+q) \setminus (\mathbf{o}(p) \oplus \mathbf{o}(q))$	$O(p+q)/(O(p) \times O(q))$
CII(−1, −1, 1)	$L = p+q$, $p = 2r, q = 2s$	$\begin{pmatrix} \mathbf{0}_{pp} & \begin{matrix} \mathbf{h}_{aa} & \mathbf{h}_{ab} \\ -\mathbf{h}_{ab}^{*} & \mathbf{h}_{aa}^{*} \end{matrix} \\ \text{h.c.} & \mathbf{0}_{qq} \end{pmatrix}$	$4rs$	$\mathbf{usp}(p+q) \setminus (\mathbf{usp}(p) \oplus \mathbf{usp}(q))$	$USp(2(r+s))/(USp(2r) \times USp(2s))$
CI(+1, −1, 1)	$L = 2M$	$\begin{pmatrix} \mathbf{0}_{MM} & \mathbf{h}_{MM} \\ \mathbf{h}_{MM}^{*} & \mathbf{0}_{MM} \end{pmatrix}$, $\mathbf{h}_{MM}^{T} = \mathbf{h}_{MM}$	$M(M+1)$	$\mathbf{usp}(2M) \setminus \mathbf{u}(M)$	$USp(2M)/U(M)$
DIII(−1, +1, 1)	$L = 2M$	$\begin{pmatrix} \mathbf{0}_{MM} & \mathbf{h}_{MM} \\ -\mathbf{h}_{MM}^{*} & \mathbf{0}_{MM} \end{pmatrix}$, $\mathbf{h}_{MM}^{T} = -\mathbf{h}_{MM}$	$M(M-1)$	$\mathbf{o}(2M) \setminus \mathbf{u}(M)$	$O(2M)/U(M)$

The classes A and AIII are complex classes, while the remainder of the classes involve a "reality condition" of the form $\mathbf{H} = \mathbf{H}^*$ and are the real classes.

All the results of this section are tabulated in the Table 2.2.

2.9 Momentum Space Representation

Since in condensed matter we invariably look at lattice based systems, it is instructive to analyze how these same symmetry conditions constraint the Hamiltonians in their momentum space representation. The generic structure of a hopping Hamiltonian is of the form

$$\mathscr{H} = \sum_{i,\delta,\alpha,\beta} \left(t_{\alpha\beta}(\delta)c^\dagger_{i,\alpha}c_{i+\delta,\beta} + \text{h.c.} \right). \tag{2.91}$$

Here, we describe hopping of a fermion between two sites in terms of $\delta \equiv (\delta_1, \delta_2 \ldots \delta_d)$ (displacement vectors) in d dimensions. $t_{\alpha\beta}(\delta)$ is the corresponding hopping integral coupling the flavors α and β at the two sites. Every site can be assumed to be made of a GFS and the complete system is a periodic arrangement of such GFS at every site in a d-dimensional volume V. $c_{i\alpha}$, $c^\dagger_{i\alpha}$ is the fermion annihilation and creation operator at site i with flavor α respectively. Using,

$$c_{i,\alpha} = \frac{1}{\sqrt{V}} \sum_k \exp(\mathrm{i}\mathbf{k}.\mathbf{r}_i)c_{k,\alpha} \tag{2.92}$$

$$c^\dagger_{i,\alpha} = \frac{1}{\sqrt{V}} \sum_k \exp(-\mathrm{i}\mathbf{k}.\mathbf{r}_i)c^\dagger_{k,\alpha} \tag{2.93}$$

Eq. (2.91) becomes,

$$\mathscr{H} = \sum_{k,\alpha\beta} H_{\alpha\beta}(k)c^\dagger_{k,\alpha}c_{k,\beta} \tag{2.94}$$

where

$$H_{\alpha\beta}(k) = \sum_\delta t_{\alpha\beta}(\delta)\exp(\mathrm{i}\mathbf{k}.\delta). \tag{2.95}$$

Therefore symmetry constraints in $\mathbf{H}$, as discussed in previous sections, will appear as constraints on $H_{\alpha\beta}(k)$. It is conditions on these which we now obtain.

Class A: There are no constraints on $H_{\alpha\beta}(k)$ in this class.

Class AI: The canonical transformation for time reversal operator in this class is $U_T = 1$ (see Table 2.1). The action on momentum space operator is

$$\mathscr{T} c_{k,\alpha} \mathscr{T}^{-1} = \mathscr{T} \frac{1}{\sqrt{V}} \sum_{k} \exp(-i\mathbf{k}.\mathbf{r}_i) c_{i,\alpha} \mathscr{T}^{-1} = \frac{1}{\sqrt{V}} \sum_{k} \exp(i\mathbf{k}.\mathbf{r}_i) c_{i,\alpha} = c_{-k,\alpha}.$$

$$(2.96)$$

Similarly, $\mathscr{T} c_{k,\alpha}^{\dagger} \mathscr{T}^{-1} = c_{-k,\alpha}^{\dagger}$. This provides the following transformation

$$\mathscr{T} \mathscr{H} \mathscr{T}^{-1} = \mathscr{T} \sum_{k,\alpha,\beta} H_{\alpha\beta}(k) c_{k,\alpha}^{\dagger} c_{k,\beta} \mathscr{T}^{-1} = \sum_{k,\alpha,\beta} H_{\alpha\beta}^*(k) c_{-k,\alpha}^{\dagger} c_{-k,\beta} = \sum_{k,\alpha,\beta} H_{\alpha\beta}^*(-k) c_{k,\alpha}^{\dagger} c_{k,\beta}.$$

$$(2.97)$$

Therefore a Hamiltonian will be in class AI if

$$H_{\alpha\beta}^*(-k) = H_{\alpha\beta}(k). \tag{2.98}$$

Class AII: This class has symmetry signature $(-1, 0, 0)$ with $\mathbf{U}_T = \mathbf{J}$ (L = even). Consider M orbitals with flavor labels α, β such that they transform into one another as

$$\begin{aligned} \mathscr{T} c_{i\alpha}^{\dagger} \mathscr{T}^{-1} &= -c_{i\beta}^{\dagger} & \mathscr{T} c_{i\beta}^{\dagger} \mathscr{T}^{-1} &= c_{i\alpha}^{\dagger} \\ \mathscr{T} c_{k,\alpha} \mathscr{T}^{-1} &= -c_{-k,\beta} & \mathscr{T} c_{k,\beta} \mathscr{T}^{-1} &= c_{-k,\alpha}. \end{aligned} \tag{2.99}$$

The action on the Hamiltonian is given by,

$$\mathscr{T} \sum_{k} \begin{bmatrix} c_{k\alpha}^{\dagger} & c_{k\beta}^{\dagger} \end{bmatrix} \begin{bmatrix} H_{\alpha\alpha}(k) & H_{\alpha\beta}(k) \\ H_{\beta\alpha}(k) & H_{\beta\beta}(k) \end{bmatrix} \begin{bmatrix} c_{k\alpha} \\ c_{k\beta} \end{bmatrix} \mathscr{T}^{-1} \tag{2.100}$$

$$= \sum_{k} \begin{bmatrix} -c_{-k\beta}^{\dagger} & c_{-k\alpha}^{\dagger} \end{bmatrix} \begin{bmatrix} H_{\alpha\alpha}(k) & H_{\alpha\beta}^*(k) \\ H_{\beta\alpha}^*(k) & H_{\beta\beta}(k) \end{bmatrix} \begin{bmatrix} -c_{-k\beta} \\ c_{-k\alpha} \end{bmatrix} \tag{2.101}$$

$$= \sum_{k} \begin{bmatrix} c_{k\alpha}^{\dagger} & c_{k\beta}^{\dagger} \end{bmatrix} \begin{bmatrix} H_{\beta\beta}(-k) & -H_{\beta\alpha}^*(-k) \\ -H_{\alpha\beta}^*(-k) & H_{\alpha\alpha}(-k) \end{bmatrix} \begin{bmatrix} c_{k\alpha} \\ c_{k\beta} \end{bmatrix}. \tag{2.102}$$

The constraints on the Hamiltonian is

$$\begin{bmatrix} H_{\alpha\alpha}(k) & H_{\alpha\beta}(k) \\ H_{\beta\alpha}(k) & H_{\beta\beta}(k) \end{bmatrix} = \begin{bmatrix} H_{\beta\beta}(-k) & -H_{\beta\alpha}^*(-k) \\ -H_{\alpha\beta}^*(-k) & H_{\alpha\alpha}(-k) \end{bmatrix}. \tag{2.103}$$

Class D: In this class $\mathbf{U}_C = \mathbf{1}$. The momentum space operators transform as

$$\mathscr{C} c_{k,\alpha} \mathscr{C}^{-1} = \mathscr{C} \frac{1}{\sqrt{V}} \sum_{k} \exp(-i\mathbf{k}.\mathbf{r}_i) c_{i,\alpha} \mathscr{C}^{-1} = \frac{1}{\sqrt{V}} \sum_{k} \exp(-i\mathbf{k}.\mathbf{r}_i) c_{i,\alpha}^{\dagger} = c_{-k,\alpha}^{\dagger}.$$

$$(2.104)$$

Similarly, $\mathscr{C} c_{k,\alpha}^\dagger \mathscr{C}^{-1} = c_{-k,\alpha}$. This implies

$$\mathscr{C}\mathscr{H}\mathscr{C}^{-1} = \mathscr{C} \sum_{k,\alpha,\beta} H_{\alpha\beta}(k) c_{k,\alpha}^\dagger c_{k,\beta} \mathscr{C}^{-1} = \sum_{k,\alpha,\beta} H_{\alpha\beta}(k) c_{-k,\alpha} c_{-k,\beta}^\dagger = -\sum_{k,\alpha,\beta} H_{\alpha\beta}^*(-k) c_{k,\alpha}^\dagger c_{k,\beta}.$$

(2.105)

In the last step we have assumed that $H_{\alpha,\beta}(k)$ is traceless. Therefore a class **D** Hamiltonian satisfies

$$- H_{\alpha\beta}^*(-k) = H_{\alpha\beta}(k). \tag{2.106}$$

Class C: This class has the symmetry signature $(0, -1, 0)$ and can again accommodate only a even number of orbitals per site. Similar to class **AII** we label the states with α, β flavors and find their transformations to be

$$\mathscr{T} c_{i\alpha}^\dagger \mathscr{T}^{-1} = -c_{i\beta} \qquad \mathscr{T} c_{i\beta}^\dagger \mathscr{T}^{-1} = c_{i\alpha}$$
$$\mathscr{T} c_{k,\alpha} \mathscr{T}^{-1} = -c_{-k,\beta}^\dagger \qquad \mathscr{T} c_{k,\beta} \mathscr{T}^{-1} = c_{-k,\alpha}^\dagger. \tag{2.107}$$

Proceeding on the similar lines as before one obtains

$$\begin{bmatrix} H_{\alpha\alpha}(k) & H_{\alpha\beta}(k) \\ H_{\beta\alpha}(k) & H_{\beta\beta}(k) \end{bmatrix} = \begin{bmatrix} -H_{\beta\beta}(-k) & H_{\alpha\beta}(-k) \\ H_{\beta\alpha}(-k) & -H_{\alpha\alpha}(-k) \end{bmatrix}. \tag{2.108}$$

where we have also assumed the Hamiltonian is traceless.

Class AIII: This class has the symmetry signature $(0, 0, 1)$. The canonical representation for $\mathbf{U}_S$ is $\mathbf{1}_{p,q}$. The states transform as

$$\mathscr{S} c_{k,\alpha} \mathscr{S}^{-1} = \mathscr{S} \frac{1}{\sqrt{V}} \sum_k \exp(-\mathrm{i} k.r_i) c_{i,\alpha} \mathscr{S}^{-1} = \frac{1}{\sqrt{V}} \sum_k \exp(\mathrm{i} k.r_i) c_{i,\alpha}^\dagger = c_{k,\alpha}.$$

(2.109)

Similarly $\mathscr{S} c_{k,\beta} \mathscr{S}^{-1} = -c_{k,\beta}$.

Therefore the condition on the Hamiltonian becomes

$$\begin{bmatrix} H_{\alpha\alpha}(k) & H_{\alpha\beta}(k) \\ H_{\beta\alpha}(k) & H_{\beta\beta}(k) \end{bmatrix} = \begin{bmatrix} -H_{\alpha\alpha}(k) & H_{\alpha\beta}(k) \\ H_{\beta\alpha}(k) & -H_{\beta\beta}(k) \end{bmatrix}. \tag{2.110}$$

This compels the Hamiltonian to have the form

$$\begin{bmatrix} 0 & H_{\alpha\beta}(k) \\ H_{\beta\alpha}(k) & 0 \end{bmatrix} \tag{2.111}$$

which gives this matrix a off-diagonal structure.

Class BDI: This is one of the symmetry classes which has all the three symmetries and a signature of $(+1, +1, 1)$. Given this has the sublattice symmetry the Hamiltonian is already constrained to be of the form given in Eq. (2.111). Presence of time reversal (with $\mathbf{U_T = 1}$) will further impose

$$H_{\alpha\beta}(k) = H_{\alpha\beta}^*(-k). \tag{2.112}$$

Class CII: This has symmetry signature $(-1, +1, 1)$. For time reversal symmetry, the site based operators transform as

$$\mathcal{T} c_{i\alpha_p} \mathcal{T}^{-1} = -c_{i\beta_p} \qquad \mathcal{T} c_{i\beta_p} \mathcal{T}^{-1} = c_{i\alpha_p}$$
$$\mathcal{T} c_{i\alpha_q} \mathcal{T}^{-1} = -c_{i\beta_q} \qquad \mathcal{T} c_{i\beta_q} \mathcal{T}^{-1} = c_{i\alpha_q}.$$

The corresponding momentum states get transformed as

$$\mathcal{T} c_{k\alpha_p} \mathcal{T}^{-1} = -c_{-k\beta_p} \qquad \mathcal{T} c_{k\beta_p} \mathcal{T}^{-1} = c_{-k\alpha_p}$$
$$\mathcal{T} c_{k\alpha_q} \mathcal{T}^{-1} = -c_{-k\beta_q} \qquad \mathcal{T} c_{k\beta_q} \mathcal{T}^{-1} = c_{-k\alpha_q}.$$

The constraint on the Hamiltonian is given by

$$\left(\begin{array}{c|cc} 0 & H_{\beta_p\beta_q}^*(-k) & -H_{\beta_p\alpha_q}^*(-k) \\ & -H_{\alpha_p\beta_q}^*(-k) & H_{\alpha_p\alpha_q}^*(-k) \\ \hline \text{h.c.} & & 0 \end{array} \right) = \left(\begin{array}{c|cc} 0 & H_{\alpha_p\alpha_q}(k) & H_{\alpha_p\beta_q}(k) \\ & H_{\beta_p\alpha_q}(k) & H_{\beta_p\beta_q}(k) \\ \hline \text{h.c.} & & 0 \end{array} \right). \tag{2.113}$$

Class CI: This class has the symmetry signature $(+1, -1, 1)$. Along with sublattice one needs to implement time reversal with $\mathbf{U_T = F}$. The time reversal gets implemented by

$$\mathcal{T} c_{i\alpha}^\dagger \mathcal{T}^{-1} = c_{i\beta}^\dagger \qquad \mathcal{T} c_{i\beta}^\dagger \mathcal{T}^{-1} = c_{i\alpha}^\dagger$$
$$\mathcal{T} c_{k,\alpha} \mathcal{T}^{-1} = c_{-k,\beta} \qquad \mathcal{T} c_{k,\beta} \mathcal{T}^{-1} = c_{-k,\alpha}. \tag{2.114}$$

The Hamiltonian constraint becomes

$$\begin{bmatrix} 0 & H_{\alpha\beta}(k) \\ H_{\beta\alpha}(k) & 0 \end{bmatrix} = \begin{bmatrix} 0 & H_{\beta\alpha}^*(-k) \\ H_{\alpha\beta}^*(-k) & 0 \end{bmatrix}. \tag{2.115}$$

Class DIII: This has the symmetry signature $(-1, +1, 1)$. The Hamiltonian structure in Eq. (2.111) gets further constrained due to time reversal with $\mathbf{U_T = J}$. The analysis follows closely the class AII and class CI and the final condition is

$$\begin{bmatrix} 0 & H_{\alpha\beta}(k) \\ H_{\beta\alpha}(k) & 0 \end{bmatrix} = \begin{bmatrix} 0 & -H_{\beta\alpha}^*(-k) \\ -H_{\alpha\beta}^*(-k) & 0 \end{bmatrix}. \tag{2.116}$$

2.10 Examples

We now present some of the model topological Hamiltonians which fall in different symmetry classes and transform them to their canonical forms.

2.10.1 BHZ Model of a Chern Insulator

Consider the following model for a Chern insulator on a square lattice [12] which was discussed in Sect. 1.4.3. The Hamiltonian is given by

$$H = \sigma_x \sin k_x + \sigma_y \sin k_y + B(M + 2 - \cos k_x - \cos k_y)\sigma_z. \qquad (2.117)$$

As we had discussed then, this model is known to have topological transitions for $B \neq 0$ and for different regimes of M. Let us now analyze the symmetries of this model. This model has parity implemented by σ_z operator $(\sigma_z H(k)\sigma_z = H(-k))$. This is one of the ordinary symmetries. The system also has charge conjugation symmetry, implemented by $\mathbf{U_C} = \mathrm{i}\sigma_x$ $(\sigma_x H(k)^*\sigma_x = -H(-k))$. Since $\sigma_x^2 = 1$, the symmetry signature is $(0, 1, 0)$ and the model belongs to class D.

However, these symmetry representations are not in their canonical forms (see Table 2.1). The canonical representation of $\mathbf{U_C}$ in this class is given by $\mathbf{1}$. Let us implement a transformation such that (see discussion near Eq. (2.46))

$$\mathbf{R1R}^T = \mathrm{i}\sigma_x. \qquad (2.118)$$

This can be achieved by choosing $\mathbf{R} = \frac{1}{\sqrt{2}}(\sigma_y + \sigma_z)$. The Hamiltonian gets transformed by

$$H(k) \rightarrow \mathbf{R}^\dagger H(k)\mathbf{R} \qquad (2.119)$$

into

$$H = -\sigma_x \sin k_x + \sigma_z \sin k_y + B(M + 2 - \cos k_x - \cos k_y)\sigma_y. \qquad (2.120)$$

This Hamiltonian satisfies $H^*(k) = -H(-k)$, the constraint for a Class D Hamiltonian (see Eq. (2.106)).

2.10.2 A Model for Topological Insulator

Let us consider another model [12, 13] where Hamiltonian is given by

$$H = \sin k_x \sigma_x \otimes s_z + \sin k_y \sigma_y \otimes \mathbf{1} + (M + 2 - \cos k_x - \cos k_y)\sigma_z \otimes \mathbf{1}$$
$$+ g(\sin k_x \sigma_x \otimes s_x + \sin k_y \sigma_x \otimes s_y). \tag{2.121}$$

The system clearly has time reversal symmetry given by $\mathbf{1} \otimes s_y \mathbf{K}$. We also identify parity by $\sigma_z \otimes \mathbf{1}$. The charge conjugation symmetry is implemented by $-i\sigma_x \otimes \frac{s_z + gs_x}{\sqrt{1+g^2}}\mathbf{K}$ and sublattice (chiral) symmetry is given by $\sigma_x \otimes \frac{gs_z - s_x}{\sqrt{1+g^2}}$. Therefore this has all the three non-ordinary symmetries and belongs to class DIII. Summarizing

$$\mathbf{U}_T : \mathbf{1} \otimes s_y \qquad \mathbf{U}_C : -i\sigma_x \otimes \frac{s_z + gs_x}{\sqrt{1+g^2}} \qquad \mathbf{U}_S : \sigma_x \otimes \frac{gs_z - s_x}{\sqrt{1+g^2}}. \tag{2.122}$$

The canonical representations for Class DIII are given by (see Table 2.1),

$$\mathbf{U}_T = \mathbf{J} = i\mathbf{1} \otimes s_y \qquad \mathbf{U}_C : \mathbf{F} = \mathbf{1} \otimes s_x \qquad \mathbf{U}_S : \mathbf{1}_{2,2} = \mathbf{1} \otimes s_z. \tag{2.123}$$

We can perform a basis change transformation such that (see Eq. (2.53))

$$\mathbf{R}^\dagger \left(\sigma_x \otimes \frac{gs_z - s_x}{\sqrt{1+g^2}}\right)\mathbf{R} = \mathbf{1}_{2,2} = \mathbf{1} \otimes s_z \tag{2.124}$$

and transform $H(\mathbf{k}) \rightarrow \mathbf{R}^\dagger H(\mathbf{k})\mathbf{R}$. Using

$$\mathbf{R} = \frac{1}{2\sqrt{2}\sqrt{1+g^2}}\Big(g(-\mathbf{1} \otimes s_x - \sigma_x \otimes s_z + i\sigma_y \otimes s_z + \sigma_z \otimes s_x)$$
$$+ (1 + \sqrt{1+g^2})(-\mathbf{1} \otimes s_z + \sigma_x \otimes s_x) + (-1 + \sqrt{1+g^2})(i\sigma_y \otimes s_x - \sigma_z \otimes s_z)\Big) \tag{2.125}$$

the transformed Hamiltonian reads

$$H = -\sqrt{1+g^2}\Big(\sin k_x(\mathbf{1} \otimes s_x) + \sin k_y(\sigma_z \otimes s_y)\Big) + (2 + M - \cos k_x - \cos k_y)\sigma_y \otimes s_y. \tag{2.126}$$

This transformed Hamiltonian satisfies all the three non-ordinary symmetries with symmetry representations in their canonical forms (see Eq. (2.123)). The Hamiltonian also has an off-diagonal structure and satisfies Eq. (2.116).

2.11 Outlook

In this chapter we have derived, from first principles the ten symmetry classes and the generic structure of the Hamiltonians in each of them. Crucial to this was the realization that the graded vector space and basic definitions of symmetry operations allows for only four types of symmetries. Next important step was to find that time reversal, charge conjugation and sublattice were in fact the three non-ordinary symmetries. The geometric structure of the time evolution operator was also found. In order to connect these results to the lattice based Hamiltonians, we have also looked at the momentum space representations. Later in the chapter we presented some model Hamiltonians, well known in literature, and transformed them to the canonical forms. This chapter provides crucial new advancements. (i) It provides a unifying framework to build the tenfold symmetry classification. (ii) The formalism allows one to set up generic Hamiltonians, not just on lattice sites, but orbital spaces and arbitrary motifs which by construction would respect various symmetries. (iii) As we will see in the coming chapters this symmetry analysis also allows us to set up topological Hamiltonians on amorphous systems and on fractals. (iv) The methods and analysis will also be at the forefront in Chap. 7 where we discuss and obtain the structure of many-body Hamiltonians in different symmetry classes.

References

1. Altland A, Zirnbauer MR (1997) Nonstandard symmetry classes in mesoscopic normal-superconducting hybrid structures. Phys Rev B 55:1142–1161
2. Heinzner P, Huckleberry A, Zirnbauer M (2005) Symmetry classes of disordered fermions. Commun Math Phys 257(3):725–771
3. Zirnbauer MR (2010) Symmetry classes. ArXiv e-prints, arXiv:1001:0722
4. Dyson FJ (1962) The threefold way. Algebraic structure of symmetry groups and ensembles in quantum mechanics. J Math Phys 3(6):1199–1215
5. Sakurai JJ, Tuan S-F, Commins ED (1995) Modern quantum mechanics, revised edn
6. Wigner EP (1959) Group theory and its application to the quantum mechanics of atomic spectra. Academic, New York
7. Parthasarathy KR (1969) Projective unitary antiunitary representations of locally compact groups. Commun Math Phys 15(4):305–328
8. Ludwig AWW (2016) Topological phases: classification of topological insulators and super-conductors of non-interacting fermions, and beyond. Phys Scripta 2016(T168):014001. arXiv:1512.08882
9. Dreiner HK, Haber HE, Martin SP (2010) Two-component spinor techniques and feynman rules for quantum field theory and supersymmetry. Phys Rep 494(12):1–196
10. Caselle M, Magnea U (2004) Random matrix theory and symmetric spaces. Phys Rep 394(23):41–156
11. Gilmore R (1974) Lie groups, Lie algebras, and some of their applications. Wiley, New York
12. Bernevig BA, Hughes TL (2013) Topological insulators and topological superconductors. Princeton University Press, Princeton
13. Shen S-Q (2013) Topological insulators: Dirac equation in condensed matters, vol 174. Springer Science & Business Media, Berlin

Chapter 3
Topological Insulators in Amorphous Systems

3.1 Introduction

In the last chapter we saw how symmetry analysis allowed us to construct Hamiltonians in different symmetry classes. We now embark upon investigating topological phases in context of *ill* systems. As we discussed in Sect. 1.4.3, initial clues to topological phases were provided by the discovery of integer quantum Hall effect [1] and the theoretical work that followed [2–4]. These ideas saw a resurgence with the discovery of the two dimensional spin Hall insulator [5–10], soon followed [11–14] by the three dimensional topological insulator (see [15–17]). A complete classification of gapped phases of noninteracting fermions then appeared [18–21]. This classification hinges on the ten symmetry classes of Altland and Zirnbauer [22] which we discussed in the last chapter. In any given spatial dimension d only five symmetry classes of the ten host topologically nontrivial phases (see e.g., Kitaev [21] and the Table 1.1). Most of the topological phases in various symmetry classes are based on clean systems, building on ideas developed on Brillouin zone etc. which we discuss below.

Two gapped systems are considered to be topologically equivalent if the ground state of one can be reached starting from the ground state of the other by a symmetry preserving adiabatic deformation of the Hamiltonian which does not close the gap during the deformation process. The ground state of a gapped crystalline system is characterized by a set of filled bands. For each point in the Brillouin zone (BZ), this amounts to a Slater determinant state made of Bloch wave functions (whose character is determined by the symmetry class) of the filled bands. In d-dimensions this turns out to be a map from the BZ (d-dimensional torus) to points in a symmetric space whose character is determined by the total number of bands and the symmetry of the system [21]. Whether such a map allows for nontrivial "winding" decides if a symmetry class supports topological phases in that dimension. Well known lattice models of Haldane [4], Kitaev [23], Kane–Mele [6], Bernevig–Hughes–Zhang (BHZ) [9], etc. fall into this paradigm. Some of this models were discussed in Sect. 1.4.3. Gapped topological phases realized in such systems are robust to impurities and

© Springer Nature Switzerland AG 2019
A. Agarwala, *Excursions in Ill-Condensed Quantum Matter*,
Springer Theses, https://doi.org/10.1007/978-3-030-21511-8_3

disorder (that preserve the symmetry), and their surfaces typically host gapless modes which open possible new directions for technological applications [15–17]. These ideas have strongly influenced the experimental search of topological systems both in material [24–27] and synthetic cold-atomic setups [28].

While robustness to disorder is the defining property of any topological state, usually it is stated with the caveat "disorder should not be large enough to close the gap". Most studies which have attempted to address the question of disorder in topological systems have, therefore, an implicit crystalline lattice and a "small" disorder. The effect of onsite (Anderson) disorder on a $d = 3$ topological insulator was studied in Ref. [29], where it was shown that with increasing disorder a topological state transits to a metal. The effect of disorder in Kitaev chains has also been studied [30]. Other pertinent studies include, how an onsite disorder can induce [31] a topological phase in a trivial system (which has the necessary ingredients to produce topological phases), and an interesting concept of "statistical topological insulators" which requires another statistical symmetry apart from the symmetry protecting the topological phase [32]. Questions related to the robustness against disorder have also been posed in the context of weak topological insulators where it has been shown that they are surprisingly robust to disorder than previously expected [33]. Physics of topological phases in noncrystalline lattices have received relatively less attention. Most studies have focused on quasicrystalline systems [34–36], for example, realizing a weak topological insulator phase in such a system [35]. An interesting unexplored question, both from theoretical and practical perspectives, is whether a completely random set of points i.e., a random lattice, such as that realized by impurities in a material, can host topological phases. This is the question that we address in this chapter. Random lattices have previously been explored in the context of lattice field theory [37], however, the possibility of topological phases in them has not been so far addressed.

In this chapter, we will theoretically establish that amorphous systems can host topologically insulating phases. We will provide a demonstration of this by constructing models (using familiar ingredients) on random lattices where fermions hop between sites within a finite range. By tuning parameters (such as the density of sites), we show that the system undergoes a quantum phase transition from a trivial to a topological phase. We characterize the topological nature by obtaining the topological invariant (cf. Bott index [38]), and associated quantized transport signatures. We also address interesting features of such quantum phase transitions. This is achieved through a detailed study of all nontrivial symmetry classes (A, AII, D, DIII and C) in two dimensions (see Table 1.1). We will also provide a demonstration of a $\mathbb{Z}_2$ topological insulator in three dimensions. This work opens a new direction in the experimental search for topological quantum matter, by demonstrating their possibility in, as yet unexplored, amorphous systems. We discuss several examples including glassy systems and other engineered random systems.

3.2 The Model

We consider a region (box) of d-dimensional space whose volume is V. Our random lattice is constructed by placing N sites (labeled by $I = 1, \ldots, N$) randomly in the box (see Fig. 3.1). The positions of the sites are sampled from an uncorrelated uniform distribution. The collection of points is characterized by a single parameter, viz.,

$$\rho = \frac{N}{V}. \tag{3.1}$$

The sites are identical, each of which hosts L single particle states. The states at the site I are denoted by $|I\alpha\rangle$ and the associated fermion operators are denoted by $c^\dagger_{I,\alpha}$ and $c_{I,\alpha}$ ($\alpha = 1, \ldots, L$). The "flavor" label α can stand for the spin quantum number or an orbital quantum number as appropriate.

Fermions can hop from site to site, possibly changing their flavor in the process. We mimic a realistic system by considering a finite range R of hopping (see Fig. 3.1). A generic Hamiltonian of such a system is

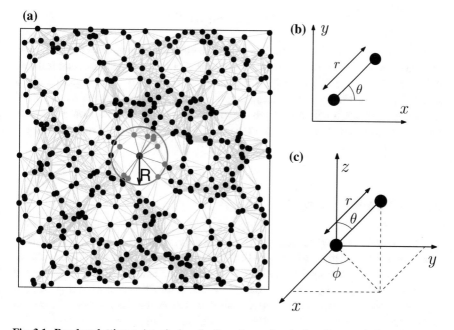

Fig. 3.1 Random lattice: **a** A typical realization of a random lattice. Sites are indicated by dark spots. Light lines indicate the hoppings between the sites which are within the distance $\leq R$ from each other. **b, c** The separation between two sites is described by a distance r and an angle θ in two dimensions (**b**) and by (θ, ϕ) in three dimensions (**c**)

$$\mathcal{H} = \sum_{I\alpha} \sum_{J\beta} t_{\alpha\beta}(\boldsymbol{r}_{IJ}) c^{\dagger}_{I,\alpha} c_{J,\beta}. \tag{3.2}$$

Here, J runs over all sites that are within a distance of R of I, i.e., $|\boldsymbol{r}_{IJ}| \leq R$ (with $\boldsymbol{r}_{IJ}$ being the vector from I to J). The $L \times L$ matrix $t_{\alpha\beta}(\boldsymbol{r})$ depends on the vector $\boldsymbol{r} = r\hat{\boldsymbol{r}}$ via

$$t_{\alpha\beta}(\boldsymbol{r}) = t(r) T_{\alpha\beta}(\hat{\boldsymbol{r}}). \tag{3.3}$$

The distance dependence of the hopping is captured by $t(r)$ and $T_{\alpha\beta}(\hat{\boldsymbol{r}})$ contains both the orbital and angular dependencies. This form is motivated by hoppings found in spin-orbit coupled systems that are common in the realization of topological phases. The unit vector $\hat{\boldsymbol{r}}$ in two spatial dimensions is described by an angle θ with respect to a chosen axis (see Fig. 3.1b), while in three dimensions we parametrize $\hat{\boldsymbol{r}}$ by polar and azimuthal angles (θ, ϕ). The hopping matrix for $\boldsymbol{r} = \boldsymbol{0}$

$$t_{\alpha\beta}(\boldsymbol{0}) = \epsilon_{\alpha\beta} \tag{3.4}$$

describes the "onsite energy" or the "atomic" Hamiltonian of the system. For, $r \neq 0$, we choose

$$t(r) = C\Theta(R - r)e^{-r/a} \tag{3.5}$$

where the constant C is chosen such that $t(r)$ is a unit energy when $r = a$, i.e., $C = e$. The step function Θ enforces the cutoff distance R. In the construction that follows, the forms of $\epsilon_{\alpha\beta}$ and $T_{\alpha\beta}(\hat{\boldsymbol{r}})$ will be motivated by systems that are experimentally relevant [9, 16]. Finally, to investigate the physics of topological edge states we will study the system with and without periodic boundary conditions on the box. In remaining discussion the scale a is set to unity, and all other lengths are measured in units of a.

3.3   Symmetry Classes

We begin the discussion of our results with two dimensional systems. As is known there are five symmetry classes (A, AII, D, DIII and C) that allow topological phases to exist in two dimensions. We wish construct Hamiltonians for each of these classes that respect all the relevant symmetries. However, symmetry conditions impose constraints on the allowed hoppings between sites, which we will discuss here. As we had seen in the last chapter, based on the three intrinsic symmetries: time reversal($\mathcal{T}$), charge conjugation($\mathcal{C}$) and sublattice($\mathcal{S}$), fermionic Hamiltonians can be classified into ten classes. These classes may or may not host topological physics depending on the spatial dimension d (see Table 1.1). The classes of interest to this work are:

- Class A: This class has a symmetry signature $(0, 0, 0)$ and contains no intrinsic symmetries. The topological classification in $d = 2$ is $\mathbb{Z}$ (set of integers).

Table 3.1 Hamiltonians: Onsite energies and hopping matrix elements in different symmetry classes in two dimensions. The size of the matrix determines the value of L in each class. The Cartan labels of each class and the set of independent parameters(par) in the Hamiltonian are shown in column 1. The model of class A is inspired by anomalous quantum Hall effect [39], AII by $\mathbb{Z}_2$ topological insulator model [9], D by $p + ip$ superconductor [39], DIII by time reversal invariant superconductor [40, 41] and C by $d + id$ superconductor [42, 43]

Class (par)	$\epsilon_{\alpha\beta}$	$T_{\alpha\beta}(\hat{\boldsymbol{r}})$
$A(\lambda, M, t_2)$	$\begin{pmatrix} 2+M & (1-i)\lambda \\ (1+i)\lambda & -(2+M) \end{pmatrix}$	$\begin{pmatrix} \frac{-1+t_2}{2} & \frac{-ie^{-i\theta}+\lambda(\sin^2\theta(1+i)-1))}{2} \\ -ie^{i\theta}+\lambda(\sin^2\theta(1-i)-1)) & \frac{1+t_2}{2} \end{pmatrix}$
$AII(\lambda, M, t_2, g)$	$\begin{pmatrix} 2+M+2t_2 & -i2\lambda & 0 & 0 \\ i2\lambda & -(2+M)+2t_2 & 0 & 0 \\ 0 & 0 & 2+M+2t_2 & i2\lambda \\ 0 & 0 & -i2\lambda & -(2+M)+2t_2 \end{pmatrix}$	$\begin{pmatrix} -\frac{1}{2}-\frac{t_2}{2} & -\frac{i}{2}e^{-i\theta}+\frac{i\lambda}{2} & 0 & -\frac{ig}{2}e^{-i\theta} \\ -\frac{i}{2}e^{i\theta}-\frac{i\lambda}{2} & \frac{1}{2}-\frac{t_2}{2} & -\frac{ig}{2}e^{-i\theta} & 0 \\ 0 & -\frac{ig}{2}e^{i\theta} & -\frac{1}{2}-\frac{t_2}{2} & \frac{i}{2}e^{-i\theta}+\frac{i\lambda}{2} \\ -\frac{ig}{2}e^{i\theta} & 0 & \frac{i}{2}e^{-i\theta}+\frac{i\lambda}{2} & \frac{1}{2}-\frac{t_2}{2} \end{pmatrix}$
$D(\mu, \Delta)$	$\begin{pmatrix} 2-\mu & 0 \\ 0 & -(2-\mu) \end{pmatrix}$	$\begin{pmatrix} -\frac{1}{2} & \Delta e^{i\theta} \\ -\Delta e^{-i\theta} & \frac{1}{2} \end{pmatrix}$
$DIII(M, g)$	$AII(\lambda = 0, t_2 = 0)$	$AII(\lambda = 0, t_2 = 0)$
$C(M)$	$\begin{pmatrix} 2+M & 0 \\ 0 & -(2+M) \end{pmatrix}$	$\begin{pmatrix} -\frac{1}{2} & -\frac{1}{2}e^{-i2\theta} \\ -\frac{1}{2}e^{i2\theta} & \frac{1}{2} \end{pmatrix}$

- Class AII: This class has a symmetry signature $(-1, 0, 0)$ and contains the time reversal symmetry. The topological classification in $d = 2$ is $\mathbb{Z}_2$.
- Class D: This class has symmetry signature $(0, 1, 0)$ and contains the charge conjugation symmetry. The topological classification in $d = 2$ is $\mathbb{Z}$ (set of integers).
- Class DIII: This class has symmetry signature $(-1, 1, 1)$ and contains both charge conjugation symmetry and time reversal symmetry. Presence of any two symmetries guarantees the third, and therefore this system also has the sublattice symmetry. The topological classification in $d = 2$ is $\mathbb{Z}_2$.
- Class C: This class has the symmetry signature $(0, -1, 0)$. This class does not contain the time reversal symmetry but the contains the charge conjugation with $\mathscr{C}^2 = -1$. The topological classification in $d = 2$ is $2\mathbb{Z}$ (set of even integers).

The hopping matrices are shown in Table 3.1 and are constructed in a way that the Hamiltonians reduce to standard crystalline models when implemented on hypercubic lattices with nearest neighbor hoppings (see caption of Table 3.1 for pertinent references). Definition of M with constant shifts are considered for the same objective. We now discuss how the imposition of these symmetries constrain the structure of Hamiltonians (see Table 3.1).

Let us first discuss the two orbital models (class A, D, C). None of these have the time reversal symmetry. Class D realizes $\mathscr{C}^2 = 1$ and class C realizes $\mathscr{C}^2 = -1$. Let the two orbitals, for any site I, be identified by $\alpha = \pm 1$, with $\bar{\alpha} = -\alpha$. For class D, $\mathscr{C}^2 = 1$ can be implemented by the following action on the fermionic operators

$$\mathscr{C} c_{I,\alpha}^{\dagger} \mathscr{C}^{-1} = c_{I,\bar{\alpha}} \qquad \mathscr{C} c_{I,\alpha} \mathscr{C}^{-1} = c_{I,\bar{\alpha}}^{\dagger}. \tag{3.6}$$

The system possesses the $\mathscr{C}$ symmetry if

$$: \mathscr{C} \mathcal{H} \mathscr{C}^{-1} := \mathcal{H}, \tag{3.7}$$

where : : denotes the normal ordering procedure needed to get all the creation operators to the left of the annihilation operators. Relation (3.7) imposes constraints on $t_{\alpha\beta}(\mathbf{r}_{IJ})$ (defined in Eq. (3.2)). We illustrate the derivation of these conditions in detail using the following example. Equation (3.7) states

$$: \mathscr{C} \sum_{\alpha,\beta,I,J} t_{\alpha\beta}(\mathbf{r}_{IJ}) c_{I,\alpha}^{\dagger} c_{J,\beta} \mathscr{C}^{-1} := \mathcal{H} \tag{3.8}$$

$$\implies \sum_{\alpha,\beta,I,J} t_{\alpha\beta}(\mathbf{r}_{IJ}) : \mathscr{C} c_{I,\alpha}^{\dagger} c_{J,\beta} \mathscr{C}^{-1} := \mathcal{H}. \tag{3.9}$$

Using Eqs. (3.6) and (3.9) becomes,

$$\sum_{\alpha,\beta,I,J} t_{\alpha\beta}(r_{IJ}) : c_{I,\bar{\alpha}} c^{\dagger}_{J,\bar{\beta}} : = \mathcal{H} \tag{3.10}$$

$$\implies \sum_{\alpha,\beta,I,J} t_{\alpha\beta}(r_{IJ})(\delta_{I,J}\delta_{\bar{\alpha},\bar{\beta}} - c^{\dagger}_{J,\bar{\beta}} c_{I,\bar{\alpha}}) = \mathcal{H}. \tag{3.11}$$

Thus the Hamiltonian is rendered in class **D** by choosing traceless $\epsilon_{\alpha\beta}$ and

$$- t_{\alpha\beta}(r_{IJ}) = t_{\bar{\beta}\bar{\alpha}}(r_{JI}) \quad (r_{IJ} \neq 0). \tag{3.12}$$

The model for class **D** satisfies these conditions and presence of t_2 term breaks this symmetry rendering it in class **A** (see Table 3.1). Implementation of $\mathscr{C}^2 = -1$ (considering that the orbital index is identified by $\alpha = \pm 1$) is given by

$$\mathscr{C} c^{\dagger}_{I,\alpha} \mathscr{C}^{-1} = \alpha c_{I,\bar{\alpha}} \qquad \mathscr{C} c_{I,\alpha} \mathscr{C}^{-1} = \alpha c^{\dagger}_{I,\bar{\alpha}}. \tag{3.13}$$

The condition (equivalent to Eq. (3.12)) now changes to

$$- t_{\alpha\beta}(r_{IJ})\alpha\beta = t_{\bar{\beta}\bar{\alpha}}(r_{JI}). \tag{3.14}$$

The class **C** Hamiltonian (see Table 3.1) satisfies these conditions.

We now discuss the four-orbital systems (**AII**, **DIII**). Both these contain time reversal symmetry. Additionally, **DIII** also has other symmetries. To discuss these systems, in addition to the orbital index denoted by greek letters α, β (which take value ± 1) we now introduce latin indices a and b which denote the spin (take values ± 1 for $\uparrow\downarrow$; $\bar{a} = -a$). (Note that in Eq. (3.2), in interest of brevity, we combine both spin and orbital indices into a single 'flavor' label). Time reversal symmetry can be implemented by

$$\mathscr{T} c^{\dagger}_{I,\alpha a} \mathscr{T}^{-1} = a c^{\dagger}_{I,\alpha\bar{a}} \qquad \mathscr{T} c_{I,\alpha a} \mathscr{T}^{-1} = a c_{I,\alpha\bar{a}}. \tag{3.15}$$

The condition one obtains on the Hamiltonian (with terms of the form $t_{\alpha\beta ab}(r_{IJ})$ $c^{\dagger}_{I,\alpha a} c_{J,\beta b}$) is

$$t^{*}_{\alpha\beta ab}(r_{IJ})ab = t_{\alpha\beta\bar{a}\bar{b}}(r_{IJ}). \tag{3.16}$$

Both the models **AII** and **DIII** satisfy these conditions. For the **DIII** model, sublattice symmetry is implemented using the transformation

$$\mathscr{S} c^{\dagger}_{I,\alpha a} \mathscr{S}^{-1} = \frac{1}{\sqrt{1+g^2}} \left(\bar{a} g c_{I,\bar{\alpha}a} + c_{I,\bar{\alpha}a} \right) \tag{3.17}$$

where g is the strength of the spin-orbit coupling as shown in Table 3.1.

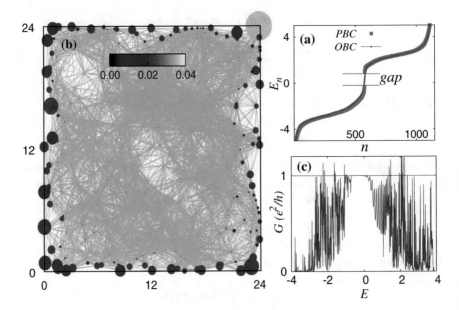

Fig. 3.2 Class A random lattice model (d=2): (Area 24×24, $R = 4$, $M = -0.5$, $t_2 = 0.25$, $\lambda = 0.5$, $\rho = 1$) **a** Energy eigenvalues E_n versus the state number n. The system with periodic boundary conditions (PBC) shows a gap; while that with open boundaries (OBC) shows mid-gap states. **b** The wave function of the mid-gap state localized on the edge. The size and the color of the blob indicates the probability of finding a fermion at that site. **c** Two terminal conductance (G) as a function of incident fermion energy showing a quantized value in the energy gap

3.4 Surprises

We have investigated the systems in all these five classes for signatures of topological phases. We now discuss the results.

3.4.1 Edge States

The class A system is realized with two orbitals per site ($L = 2$), and a Hamiltonian characterized by three dimensionless parameters λ, t_2 and M (see Table 3.1). This structure can be obtained by starting from the BHZ model [9, 39] by introducing t_2 which breaks charge conjugation symmetry. M (mass parameter) is the parameter we tune to investigate the possibility of topological phases for various values of ρ, the density of sites. We study filling of one fermion per site. Results of a particular realization of the random lattice for a 24×24 system with $\rho = 1$, $R = 4$ and $M = -0.5$ is shown in Fig. 3.2. Figure 3.2a shows the energy eigenvalues of the system with and without the periodic boundary conditions. In the presence of periodic boundary

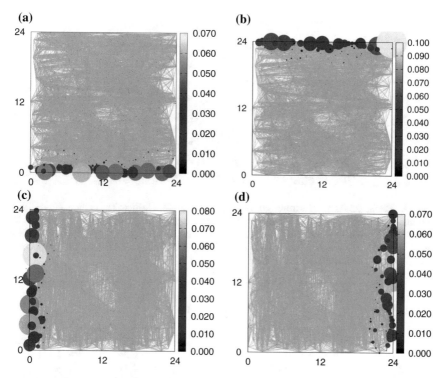

Fig. 3.3 Different boundary conditions on the class A model: (Area 24 × 24, $R = 4$, $M = -0.5$, $t_2 = 0.25$, $\lambda = 0.5$, $\rho = 1$) **a-b** The wave function of two mid-gap states localized on bottom(**a**) and top(**b**) edge when the periodic boundary conditions are applied in the horizontal direction while the vertical direction is kept under open boundary conditions. The size and the color of the blob indicates the probability of finding a fermion at that site. **c-d** Similarly, when periodic boundary conditions are applied in the vertical direction and the horizontal direction is under open boundary conditions, two typical mid gap states with wave function localized on right(**c**) and left(**d**) of the random lattice can be seen

conditions the system clearly shows an energy gap i.e, it is an insulator. In the absence of periodic boundary conditions (in both directions) we see a set of energy eigenvalues in the mid-gap region. A typical mid-gap state, we find, is an "edge" state which is localized on the "surfaces" of the box(see Fig. 3.2b) In fact, we can now impose periodic boundary conditions only in x (and keep the y direction as open) and vice-versa and show the results in Fig. 3.3. As can be seen that depending on the boundary conditions on the system, for the same parameters, one can have a state localized on any of the edges.

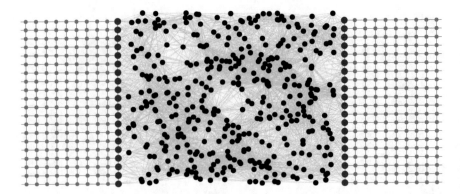

Fig. 3.4 The setup for two terminal conductance of a random lattice: The random lattice (black dots and light lines) is connected to square lattice leads (small orange sites) on both right and left sides. The square lattice can be considered as a repeating unit cell (of a single chain of atoms in vertical direction). For smooth connection of the leads to the lattice a single unit cell is appended to the device on both right and left of the random lattice (big red dots)

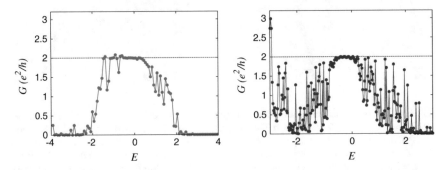

Fig. 3.5 Two terminal conductance for class C and All: (Left) The two terminal conductance is shown as a function of incident fermion energy for a single random lattice configuration for the class C model($M = -1$, $V = 60 \times 60$, $\rho = 1$, $R = 4$). The plateau at $2e^2/h$ can be clearly seen. We find the Bott index to be $= -2$. (Right) The two terminal conductance is shown as a function of incident fermion energy for a single random lattice configuration in class All ($M = -1$, $V = 16 \times 16$, $\rho = 1$, $R = 4$, $t_2 = 0.15$, $\lambda = 0.25$, $g = 0.1$). The plateau at $2e^2/h$ can also be seen for this case. Note that, unlike the class C model, this model has time reversal symmetry

3.4.2 Transport

Next to investigate the transport due to this edge state, we couple the system to leads along opposite surfaces while the other two were open. The representative setup is shown in Fig. 3.4 (Further details of implementation are provided in Appendix A.1). Figure 3.2c shows the two terminal conductance that we calculate using the non-equilibrium Green's function formalism [44]. Remarkably, when the energy of the incoming fermions is in the "bulk gap" we see the conductance is quantized to unity. Some other representative cases for other classes are shown in Fig. 3.5.

3.4.3 Bott Index

Are the edge states that we see of topological origin? The presence of an edge state that provides for a quantized conductance in a completely random lattice, if surprising, is suggestive of its topological origin. To confirm this we calculate a topological invariant, the Bott index, adapting the work of Loring and Hastings [38]. Briefly, a nontrivial Bott index quantifies the obstruction to the construction of localized Wannier orbitals from the occupied states [38]. The Bott index is given by $\frac{1}{2\pi}$ Im$\{\text{tr}(\log(WUW^{\dagger}U^{\dagger}))\}$ where U and W are obtained as follows. The position coordinates of all the orbitals $|I\alpha\rangle \equiv (x_{I\alpha}, y_{I\alpha})$ are rescaled into two angles $(\theta_{I\alpha}, \phi_{I\alpha})$ defined between $(0, 2\pi)$. Diagonal matrices Θ and Φ are defined with elements $\theta_{I\alpha}$ and $\phi_{I\alpha}$ as diagonal elements. The matrices U and W are obtained as $U = P\exp(i\Theta)P$ and $W = P\exp(i\Phi)P$ where P is the projector to the occupied states [38]. In a crystalline system, the Bott index is same as TKNN invariant [3] of 2D class A (details are discussed in Appendix A.2). Given this correspondence, it is natural to expect robust edge states when the Bott index is nontrivial. For the parameters of Fig. 3.2, we find Bott index to be -1, a nontrivial value. Taken together, the presence of bulk gap, surface localized mid-gap states, quantized transport and a nontrivial Bott index unequivocally demonstrates the topological character of this amorphous system. Remarkably the system shows a Bott index of negative unity confirming the topological character of the ground state.

3.5 Amorphous Topological Phase

Under what condition does a random lattice show topological phases? We address this question by obtaining a phase diagram in the $M - \rho$ plane, i.e., by varying the mass parameter and the density of sites. Figure 3.6a shows the Bott index for a particular configuration as a function of the mass parameter M at $\rho = 0.6$. For $-2 \lesssim M \lesssim 1.2$ the system is in a topological phase with two quantum phase transitions at $M \approx -2$ and at $M \approx 1.2$. To obtain the phase diagram we average over hundreds of realizations of the random lattice. Figure 3.6b shows a contour plot the Bott index in the $M - \rho$ plane. We see that there is a large regime of parameters in the density and M where the system is topological. An important point is that a critical density ρ_c is needed to obtain such a topological phase. The existence of such a critical density is expected to be "universal", although its precise value will be determined by the specific microscopic parameters. For any $\rho > \rho_c$, note that there are two values of M at which the system has a phase transition. To investigate the nature of these phase transitions we studied the scaling properties of the gap and the Bott index as a function of the system size.

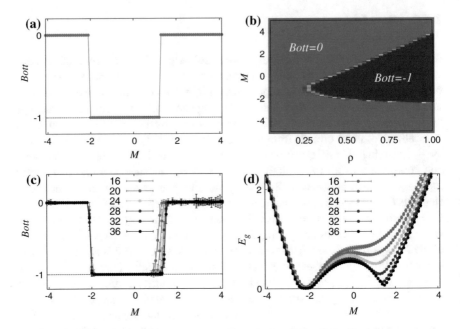

Fig. 3.6 Phase diagram: a Bott index for a particular realization of the random lattice at $\rho = 0.6$. (Area 24×24, $R = 4$, $t_2 = 0.25$, $\lambda = 0.5$) **b** Contour plot of (configuration averaged) Bott index in the $M - \rho$ plane. Red region indicates topologically nontrivial phase. **c** Configuration averaged Bott index for various system sizes. **d** Configuration averaged energy gap E_g for various system sizes. **c** and **d** are also for $\rho = 0.6$. Configuration average is performed over 320 realizations of the random lattice. Other parameters are kept same as in (**a**)

3.5.1   Finite Size Scaling

Upon reducing M from a positive value, in the first transition that is encountered, the gap does not vanish in a system of a finite size, and indeed the smallest gap as a function of M reduces with increasing system size. Indeed this is also indicated in the jump of the Bott index which becomes increasingly sharper with increasing system size. The story is quite different at the second transition that occurs at a negative M. Here the gap is very small and is less sensitive to increase in system size. Similarly the sharpness in jump of the Bott index is less sensitive to system size. This is also seen for other classes and for other densities. Some further representative cases are shown in Figs. 3.7 and 3.8.

In order to further analyze how the gap reduces with increasing system size (for the transitions occurring at positive and negative M), we find the minimum value of the gap ($\equiv \Delta E_g$) at any given V and investigate its scaling with increasing V (see Fig. 3.9). The value of the energy gap (for the transition occurring at negative M) falls as $\sim 1/V$ (see Fig. 3.9). At positive M, while the overall trend of gap closing is $\sim \frac{1}{\sqrt{V}}$, there are oscillations with system size. These oscillations are reminiscent of

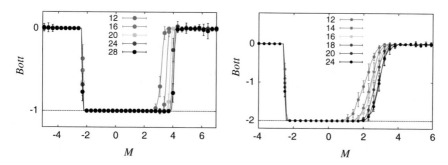

Fig. 3.7 Bott indices: (Left) Configuration averaged Bott index for various system sizes for a class A model ($R = 4$, $t_2 = 0.25$, $\lambda = 0.5$, $\rho = 1$). (Right) Similar plot for a class C model ($\rho = 1$, $R = 4$). Note that unlike class A, where the Bott index takes a nontrivial value of -1, class C has a nontrivial Bott index of -2. For (Left) configuration averaging is performed over 320 realizations of the random lattice and for (Right) it is performed over 160 realizations

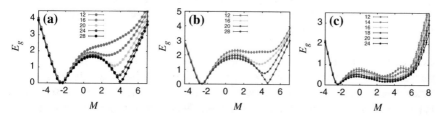

Fig. 3.8 Two kinds of transitions: Configuration averaged energy gap E_g for various system sizes for a class A model ($R = 4$, $t_2 = 0.25$, $\lambda = 0.5$, $\rho = 1.0$) **b** Similar plot for a class AII model with ($\rho = 1$, $R = 4$, $t_2 = 0.15$, $\lambda = 0.25$, $g = 0.1$), and **c** for the class C model ($\rho = 1$, $R = 4$). Note that for all the systems one of the "gap closings" (at negative M) is less sensitive to changes in the system size than the other (at positive M). For **a-b** configuration averaging is performed over 320 realizations of the random lattice and for **c** it is performed over 160 configurations

the interference effects between edge states in a strip of topological insulator [45]. We conjecture that these effects may be due to possible rare regions in the random lattice. The wave functions localized on the rare regions in different periodic images may overlap to produce this oscillating gap.

3.6 Nature of Bulk States

In Fig. 3.2 we analyzed a class A system and had seen that a typical mid-gap state lies on the outer edge of the random lattice. It is particularly interesting to analyze the nature of bulk states in the system. For this purpose we calculate the inverse participation ration (IPR) of the wave functions. For a generic wave function $|\psi\rangle = \sum_{I\alpha} \psi_{I\alpha} |I\alpha\rangle$, the IPR is given by $\sum_{I\alpha} |\psi_{I\alpha}|^4$. A low value of IPR which falls as $\sim \frac{1}{V}$, where V is the "volume" of the system, indicates a delocalized state extended

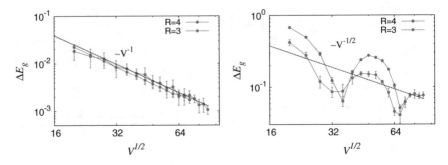

Fig. 3.9 Finite size scaling of the gap: Configuration averaged minimum energy gap ΔE_g and its scaling with various system sizes at the two topological transitions: (Left) for negative M and (Right) for positive M (class A model, $R = 4$, $t_2 = 0.25$, $\lambda = 0.5$, $\rho = 0.6$). In (Left) the gap falls as $1/V$ (shown by dashed line). For (Right) the gap falls as $1/\sqrt{V}$ (shown by a dashed line), but there are pronounced oscillations. Configuration averaging is performed over 320 realizations of the random lattice. Similar physics is seen even if we change the value of hopping range (R) keeping the other parameters same (Configuration averaging is performed over 16 realizations, $R = 3$)

over the full system. On the other hand, a high value of IPR (compared to $\frac{1}{V}$) which remains insensitive to increase in system size is used to diagnose a localized state. Since a edge state is extended only around the boundary, its IPR is expected to behave as $1/\sqrt{V}$ with increasing system size. The results for the amorphous system are shown in Fig. 3.10. We are at a parameter regime when the system is topological and has edge states. For $V = 24 \times 24$ ($L = 2$) an extended state delocalized over the complete bulk is expected to have an IPR between 10^{-3} and 10^{-4} and an edge state is expected to have an IPR between 10^{-3} and 10^{-2}. While the edge states near $E = 0$ (see Fig. 3.10a) indeed have IPR in the expected regime, the amorphous bulk states here have much larger value than expected of a delocalized bulk state, a difference which increases with increasing system size. We analyze the scaling of IPR with increasing system size at two values of energy, one at band center (cut I in Fig. 3.10a) and the other at $E = -3$ (cut II in Fig. 3.10a). In the former the states at the particular energy correspond to the edge states while in latter they correspond to the bulk states. Figure 3.10b-c shows that the edge states IPR scale expectedly as $\sim \frac{1}{\sqrt{V}}$, while bulk states IPR do not scale as $\sim \frac{1}{V}$. The wave function of a typical bulk state is shown in Fig. 3.10d. The key result of this analysis is that the only extended states in this system are the edge states.

3.7 Three Dimensional System

Given the vast interest enjoyed by the three dimensional topological insulator (a system with a $\mathbb{Z}_2$ invariant), we also investigate the possibility of realizing this in a random lattice. To this end, we consider a system with four orbitals ($L = 4$)

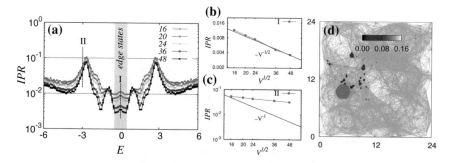

Fig. 3.10 Nature of bulk and edge states: a A configuration averaged IPR as a function of energy E for the Class A random lattice model (d=2) (Area $V = 24 \times 24$, $R = 4$, $M = -0.5$, $t_2 = \lambda = 0.0$, $\rho = 1$) with open boundary condition and in the topological regime. Close to $E = 0$ the shaded region depicts the edge states while the others are bulk states. **b** The IPR values as a function of system size along the cut I in (**a**). **c** Similar plot for the cut II in (**a**). For I, increasing system size reduces the IPR as $\frac{1}{\sqrt{V}}$ depicting that these are extended over the edge while for II, IPR do not scale as $\sim 1/V$ indicating they are *not* delocalized over the system. Moreover the IPR of bulk states are orders of magnitude higher than value expected from a state which is completely delocalized in the bulk (see text) (Configuration averaging is performed over 320 realizations for $\sqrt{V} = 16 - 36$ and 40 realizations for $\sqrt{V} = 48$). **d** The wave function of a typical bulk state for the 24×24 system (for the configuration and parameters discussed in Fig. 3.2) showing that it is localized in the bulk

with a Hamiltonian described by $\epsilon_{\alpha\beta} = \mathrm{Diag}(-3 + M, -3 + M, 3 - M, 3 - M)$ [46] where M is the mass parameter, and

$$T_{\alpha\beta}(\theta, \phi) = \frac{1}{2} \begin{pmatrix} 1 & 0 & -i\cos\theta & -ie^{-i\phi}\sin\theta \\ 0 & 1 & -ie^{i\phi}\sin\theta & i\cos\theta \\ -i\cos\theta & -ie^{-i\phi}\sin\theta & -1 & 0 \\ -ie^{i\phi}\sin\theta & i\cos\theta & 0 & -1 \end{pmatrix} \qquad (3.18)$$

This system has the required time reversal symmetry. For an appropriate set of parameters ($M = 0.5$ and $\rho = 0.6$), we find indeed that there are mid-gap states in an open system whose wave functions are localized on the boundary (see Fig. 3.11). This strongly indicates the realization of a topological state.

3.8 Perspective

The possibility of topological phases in a completely random system opens up several avenues both from experimental and theoretical perspectives. Our results suggest some new routes to the laboratory realization of topological phases. First, two dimensional systems can be made by choosing an insulating surface on which suitable "motifs" (atoms/molecules) with appropriate orbitals are deposited at random (note that this process will require far less control than conventional layered materials).

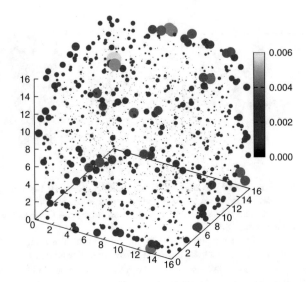

Fig. 3.11 $\mathbb{Z}_2$ **system in three dimensions**: The mid-gap state localized on the surface. The size and the color of the blob indicate the probability of finding a fermion at the site. ($\rho = 0.6$, $V = 16^3$, $M = 0.5$, $R = 4$)

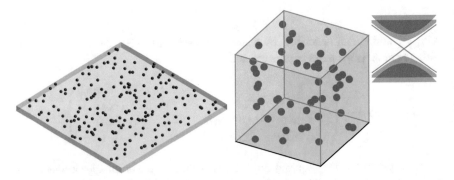

Fig. 3.12 **Perspective**: Schematic figure for both the two dimensional and three dimensional examples of realizing topological phases in random systems. In (Left) appropriate motifs can be adsorbed on a surface to produce topological states. In (Right) a large band gap insulator can be impurity doped to produce topologically protected states (see text)

The electronic states of these motifs will then hybridize to produce the required topological phase. Second is the possibility of creating three dimensional systems starting from a suitable large band gap trivial insulator. The idea then is to place "impurity atoms", again with suitable orbitals and "friendly" chemistry with the host, not unlike the process of δ-doping of phosphorus in silicon [47]. The hybridization of the impurity orbitals would again produce a topological insulating state in the impurity bands under favorable conditions (see Fig. 3.12). Here we do not thoroughly explore the specific material system that will realize these ideas. In realistic systems the temperature scales over which one will see the topological physics (determined by the bandwidth and band gap) may be low. Nevertheless, we believe that our results will motivate materials science experts to address these challenges. Finally, our work also suggests that a completely amorphous system can be a topological insulator, clearly providing new opportunities to search for "glassy" insulators with spin-orbit coupled motifs to realize topological phases.

Our work also provides some interesting new directions for theoretical research. There are two (equivalent) ways to view topological phases. The first one is "kinematic", i.e., based on the homotopy of ground state wave functions of systems in a given symmetry class as discussed in the introduction. The second approach [20], probably better suited for the random system, is based on asking if the $d-1$ dimensional surface of a gapped d-dimensional system resists localization. This absence of localization on the $d-1$ surface can be used to characterize the topology of the d dimensional bulk. Localization is prevented by the presence of a topological term in the action (nonlinear σ-model) that describes the low energy modes of the $d-1$ dimensional surface. While one usually writes down such σ-models based on symmetry considerations, an interesting question in the current context would be to uncover how such topological terms can arise in the random lattice setting.

References

1. Klitzing KV, Dorda G, Pepper M (1980) New method for high-accuracy determination of the fine-structure constant based on quantized hall resistance. Phys Rev Lett 45:494–497
2. Laughlin RB (1981) Quantized hall conductivity in two dimensions. Phys Rev B 23:5632–5633
3. Thouless DJ, Kohmoto M, Nightingale MP, den Nijs M (1982) Quantized hall conductance in a two-dimensional periodic potential. Phys Rev Lett 49:405–408
4. Haldane FDM (1988) Model for a quantum hall effect without Landau levels: condensed-matter realization of the "parity anomaly". Phys Rev Lett 61:2015–2018
5. Murakami S, Nagaosa N, Zhang S-C (2004) Spin-hall insulator. Phys Rev Lett 93:156804
6. Kane CL, Mele EJ (2005) Z_2 topological order and the quantum spin hall effect. Phys Rev Lett 95:146802
7. Kane CL, Mele EJ (2005) Quantum spin hall effect in graphene. Phys Rev Lett 95:226801
8. Bernevig BA, Zhang S-C (2006) Quantum spin hall effect. Phys Rev Lett 96:106802
9. Bernevig BA, Hughes TL, Zhang S-C (2006) Quantum spin hall effect and topological phase transition in HgTe quantum wells. Science 314(5806):1757–1761
10. König M, Wiedmann S, Brne C, Roth A, Buhmann H, Molenkamp LW, Qi X-L, Zhang S-C (2007) Quantum spin hall insulator state in HgTe quantum wells. Science 318(5851):766–770
11. Fu L, Kane CL, Mele EJ (2007) Topological insulators in three dimensions. Phys Rev Lett 98:106803
12. Moore JE, Balents L (2007) Topological invariants of time-reversal-invariant band structures. Phys Rev B 75:121306
13. Roy R (2009) Topological phases and the quantum spin hall effect in three dimensions. Phys Rev B 79:195322
14. Hsieh D, Qian D, Wray L, Xia Y, Hor YS, Cava RJ, Hasan MZ (2008) A topological dirac insulator in a quantum spin hall phase. Nature 452:970–974
15. Hasan MZ, Kane CL (2010) Colloquium: topological insulators. Rev Mod Phys 82:3045–3067
16. Qi X-L, Zhang S-C (2011) Topological insulators and superconductors. Rev Mod Phys 83(4):1057

17. Ando Y (2013) Topological insulator materials. J Phys Soc Jpn 82(10):102001
18. Qi X-L, Hughes TL, Zhang S-C (2008) Topological field theory of time-reversal invariant insulators. Phys Rev B 78:195424
19. Schnyder AP, Ryu S, Furusaki A, Ludwig AWW (2008) Classification of topological insulators and superconductors in three spatial dimensions. Phys Rev B 78:195125
20. Ryu S, Schnyder AP, Furusaki A, Ludwig AWW (2010) Topological insulators and superconductors: tenfold way and dimensional hierarchy. New J Phys 12(6):065010
21. Kitaev A (2009) Periodic table for topological insulators and superconductors. AIP Conf Proc 1134(1):22–30. http://aip.scitation.org/doi/pdf/10.1063/1.3149495
22. Altland A, Zirnbauer MR (1997) Nonstandard symmetry classes in mesoscopic normal-superconducting hybrid structures. Phys Rev B 55:1142–1161
23. Kitaev AY (2001) Unpaired majorana fermions in quantum wires. Phys Uspekhi 44(10S):131
24. Chadov S, Qi X, Kübler J, Fecher GH, Felser C, Zhang SC (2010) Tunable multi-functional topological insulators in ternary heusler compounds. Nat Mater 9(7):541–545
25. Das A, Ronen Y, Most Y, Oreg Y, Heiblum M, Shtrikman H (2012) Zero-bias peaks and splitting in an Al-InAs nanowire topological superconductor as a signature of majorana fermions. Nat Phys 8(12):887–895
26. Chang C-Z, Zhang J, Feng X, Shen J, Zhang Z, Guo M, Li K, Ou Y, Wei P, Wang L-L et al (2013) Experimental observation of the quantum anomalous hall effect in a magnetic topological insulator. Science 340(6129):167–170
27. Nadj-Perge S, Drozdov IK, Li J, Chen H, Jeon S, Seo J, MacDonald AH, Bernevig BA, Yazdani A (2014) Observation of majorana fermions in ferromagnetic atomic chains on a superconductor. Science 346(6209):602–607
28. Jotzu G, Messer M, Desbuquois R, Lebrat M, Uehlinger T, Greif D, Esslinger T (2014) Experimental realization of the topological haldane model with ultracold fermions. Nature 515(7526):237–240
29. Kobayashi K, Ohtsuki T, Imura K-I (2013) Disordered weak and strong topological insulators. Phys Rev Lett 110:236803
30. Diez M, Fulga IC, Pikulin DI, TworzydÅo J, Beenakker CWJ (2014) Bimodal conductance distribution of Kitaev edge modes in topological superconductors. New J Phys 16(6):063049
31. Li J, Chu R-L, Jain JK, Shen S-Q (2009) Topological Anderson insulator. Phys Rev Lett 102:136806
32. Fulga IC, van Heck B, Edge JM, Akhmerov AR (2014) Statistical topological insulators. Phys Rev B 89:155424
33. Ringel Z, Kraus YE, Stern A (2012) Strong side of weak topological insulators. Phys Rev B 86:045102
34. Kraus YE, Lahini Y, Ringel Z, Verbin M, Zilberberg O (2012) Topological states and adiabatic pumping in quasicrystals. Phys Rev Lett 109:106402
35. Fulga IC, Pikulin DI, Loring TA (2016) Aperiodic weak topological superconductors. Phys Rev Lett 116:257002
36. Bandres MA, Rechtsman MC, Segev M (2016) Topological photonic quasicrystals: fractal topological spectrum and protected transport. Phys Rev X 6:011016
37. Christ N, Friedberg R, Lee T (1982) Random lattice field theory: general formulation. Nucl Phys B 202(1):89–125
38. Loring TA, Hastings MB (2010) Disordered topological insulators via C*-algebras. Eur Phys Lett 92(6):67004
39. Bernevig BA, Hughes TL (2013) Topological insulators and topological superconductors. Princeton University Press, Princeton
40. Roy R (2006) Topological invariants of time reversal invariant superconductors. arXiv:cond-mat/0608064
41. Qi X-L, Hughes TL, Raghu S, Zhang S-C (2009) Time-reversal-invariant topological superconductors and superfluids in two and three dimensions. Phys Rev Lett 102:187001
42. Senthil T, Marston JB, Fisher MPA (1999) Spin quantum hall effect in unconventional superconductors. Phys Rev B 60:4245–4254

43. Chern T (2016) $d + id$ and d wave topological superconductors and new mechanisms for bulk boundary correspondence. AIP Advances 6(8)
44. Datta S (1997) Electronic transport in mesoscopic systems. Cambridge University Press, Cambridge
45. Medhi A, Shenoy VB (2012) Continuum theory of edge states of topological insulators: variational principle and boundary conditions. J Phys Condens Matter 24(35):355001
46. Fradkin E (2013) Field theories of condensed matter physics. Cambridge University Press, Cambridge
47. Scappucci G, Capellini G, Lee WCT, Simmons MY (2009) Ultradense phosphorus in germanium delta-doped layers. Appl Phys Lett 94(16):162106

Chapter 4
Seeking Topological Phases in Fractals

4.1 Introduction

As we had seen in Table 1.1, the classification of topological systems is described by a tenfold scheme and a periodicity in spatial dimensions d. Notions of integral dimensions and bulk-boundary correspondence lies at the heart of the topological band theory [1–4]. A nontrivial invariant calculated for a periodic system signals existence of robust boundary states for the same system with a boundary. This correspondence is the progenitor of formulations of various invariants such as TKNN invariant (Chern number) [5], the Pfaffian and others (for a recent review see [3]) which lead to exotic boundary physics. However, not every system has a well defined "bulk" or "boundary". Neither does every system have a well defined dimension. Is there a notion of a topological state in such systems? If yes, how can they be characterized?

Fractals [7] are systems which do not have an integral dimension and in fact, their "dimension" is characterized by the Hausdorff dimension. Hausdorff dimension quantifies how the fractal fills up space as the basic length unit is scaled. For regular systems such as lattices, Hausdorff dimension equals the usual dimension. These systems have always intrigued condensed matter physicists—extensive studies of critical phenomena have been investigated on these systems [8] and their relation to percolation problem was identified [9]. Studies have looked at solving Schrodinger operator on such lattices [10] and concept of "fractons" was introduced [11]. Further in a detailed study, analytical results were provided for the exact solution of a tight binding model on a fractal [6], showing that the spectrum also has a self-similar pattern. A recent study has shown that the transmission properties from such a lattice can capture Hausdorff dimension [12]. Interestingly, the effect of magnetic field was also investigated [13]. Spectral analysis has been conducted for Sierpinski gasket [14], a fractal, and further localization and extended states were investigated in finite and infinite Sierpinski gaskets [15–17]. Quite encouragingly, some of these structures have now been experimentally realized [18, 19]. These studies demonstrate that fractals have in fact been important building blocks of crucial insights into many physical phenomena. However, investigations pertaining to topological physics in

© Springer Nature Switzerland AG 2019
A. Agarwala, *Excursions in Ill-Condensed Quantum Matter*,
Springer Theses, https://doi.org/10.1007/978-3-030-21511-8_4

Fig. 4.1 The Sierpinski gasket: The Sierpinski gasket is constructed by recursively scooping out a fourth of a triangle. Identification of A sites and B sites makes a "periodic" system where all the sites now have the same coordination number. N is the total number of sites. The eigenvalues as a function of normalized eigenvalue number for a tight binding model defined on such a system. The result shown is for $g = 5$. In infinite g limit the spectrum has gaps and is self similar [6]

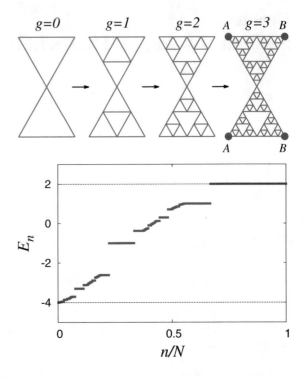

such systems have not been undertaken, perhaps because the tenets of topological physics rely heavily on notions of a well defined dimension. In one study [20] it was shown that a edge state is stable if the edge of the lattice is converted to a Koch curve.

4.2   Sierpinski Gasket

In this chapter we will investigate the topological physics in the context of the Sierpinski gasket. The standard Sierpinski gasket starts with an up facing triangle. The sides are bisected with a set of midpoints. Each of these mid points are connected to each other making new triangles. In the mean time the length of the sides is doubled. This process is now repeated and mid points are generated for all up facing triangles in every generation. This provides an object which has a Hausdorff dimension of $\frac{\log 3}{\log 2} = 1.58$. Two such Sierpinski gaskets can be connected by identifying their vertices to obtain an object which has the same dimension, but is now periodic—in the sense that each site is equivalently coordinated (introduced in [6]). In the case of the Sierpinski gasket, the coordination number of any site is 4 (see Fig. 4.1). At any generation (g) the number of distinct sites in this system grows as 3^{g+1}. The thermodynamic limit in such a system will therefore be identified by $g \to \infty$. The spectrum for this system with a tight binding model, where spinless fermions hop between every connected site with strength $-t$ $(t = 1)$ was analyzed in [6]. The spectrum is shown in Fig. 4.1. It was shown that this spectrum is self-similar, and has a number of band gaps in the thermodynamic limit. It is interesting to contrast this with the lattice based systems. Every lattice, with simple tight binding hopping and with equal coordination number at every site is gapless at all energies. A hybercubic lattice with four bonds per site will be a square lattice. For a regular lattice, number of bulk states scale as V, where V is the "volume", while the edge scales as "area". In the Sierpinski fractal, it is not apriori clear what is "edge" and what is the "bulk". For any generation g, only the sites added in the last generation can be considered as "bulk". All sites otherwise are part of one or the other "edges".

We now construct a topological Hamiltonian on the Sierpinski gasket. For this we consider a two-orbital model on each site, and the Hamiltonian is given by

$$\mathcal{H} = \sum_{I\alpha} \sum_{J\beta} t_{\alpha\beta}(r_{IJ}) c^{\dagger}_{I,\alpha} c_{J,\beta}, \tag{4.1}$$

where I, J are summed over the connected sites as shown in Fig. 4.1. $c^{\dagger}_{I\alpha}$ represents a fermion creation operator at site I with a orbital flavor α. r_{IJ} is the vector connecting sites I and J and is modeled by a basic length scale a and an angle θ. The atomic energy i.e., when $r = 0$, is given by $\epsilon_{\alpha\beta} = \text{Diag}\{2 + M, -(2 + M)\}$ and $t_{\alpha\beta}(r) = \begin{pmatrix} \frac{-1}{2} & \frac{-ie^{-i\theta}}{2} \\ \frac{-ie^{i\theta}}{2} & \frac{1}{2} \end{pmatrix}$. This model when defined on a square lattice with nearest neighbor hoppings is known to be a topological insulator in the regime $-4 < M < 0$ [21]. When defined on a triangular lattice, it is again known to be topological in $-3.5 < M < -1$ (see Appendix A.2.2). For both these lattice models, the topological transitions coincide with band gap closings and band inversion. This model was also shown to be topological on an amorphous system in the last chapter. It is important to note that for lattice based systems, the bulk band is continuous over the Brillouin zone; the concepts of Pancharatnam–Berry phase and a corresponding non-trivial Chern number, defined as an adiabatic integral over the band Berry curvature is well defined.

We investigate the system when we are at half filling (one fermion per site). The results are shown in Fig. 4.2. In Fig. 4.2a, at large positive and large negative values of M, one expects the system to be gapped (as is seen). However as M is tuned one finds a regime of gapless phase. This should be contrasted with the triangular lattice case (shown in dashed line). The gapless phase remains as the generation number is increased implying that this is a thermodynamic phase. What is the nature of the states making this gapless phase? This is shown in Fig. 4.2b. We investigate the spectrum at $M = -1$, when the system is in a gapless regime. One notices that the spectrum is reminiscent of topological systems. Some typical wave functions close to the center of the spectrum are shown. One notices that the states reside on a variety of "edges". In fact this system (as $g \to \infty$) has an infinite number of 'edge' states. This edge state spectrum is also self similar.

Fig. 4.2 "Topological" Sierpinski gasket: a At half filling the energy gap between the two central eigenvalues for various generations as a function of M is shown. The system goes through a gapless phase in an intermediate regime of M. This should be contrasted with the case when the model is built on a triangular lattice (shown by dashed line). The latter goes through a topologically insulating phase. In **b** eigenvalues and few typical "edge states" for $g = 5, M = -1$. The states remarkably resemble edge states in usual systems, but can reside on any of the "edges"

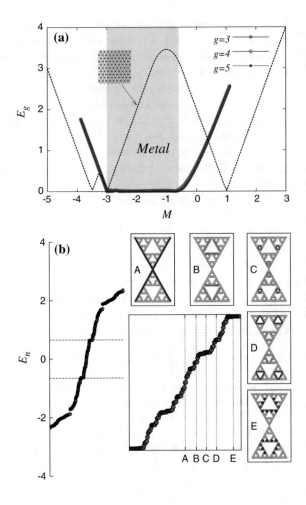

In order to analyze the complete edge spectrum we number the edges using K, where $K = 1$ shows the outermost edge and K progressively increases through integers as one goes deeper into the fractal (an example is shown in Fig. 4.3). For any wave function $|\psi\rangle = \sum \psi_i |i\rangle$, one can evaluate the overlap with the edge,

$$O_I = \sum_{i \in K} |\psi_i|^2 \qquad (4.2)$$

where sites i is summed over all the sites which belong to the edge K.

Now for any mth wave function, we define $O^m = \max\{O_K, K \in 1, \ldots, m\}$ which shows the value of the maximum probability of a wave function to be an edge state. We can also define $K^m = \{K, K \hat{=} O^m\}$ (i.e., the edge number that corresponds to the maximum overlap O_K). Therefore a state which resides, dominantly on a edge with

Fig. 4.3 Self-similar edge spectrum: (Top) The various "edges" of the Sierpinski gasket is shown ($K = 1, \ldots, 5$). (Generation(g) $= 5$) (Bottom) The states and in which edge they lie, as a function of generation number g and energy E. These are dominantly "edge" states and exist at various edges in an interesting self-similar pattern

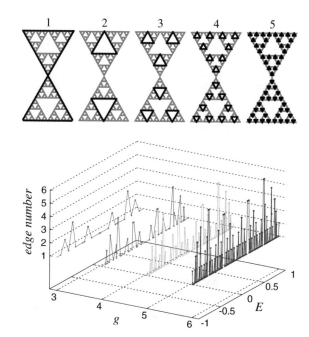

a value of O^m close to ~ 1 demonstrates that the particular wave function resides on that particular edge. The plot of K^m as a function of E and generation number g is plotted in Fig. 4.3. The values of $O_K \sim 1$ for all these states. The self similarity in the spectrum as a function of generation is manifest.

However, the "bulk" bands in this system have thermodynamic gaps. In lattices, for this model, there exists a single bulk band for $E > 0$ (and symmetrically placed $E < 0$ band). In the case of Sierpinski gasket the system has thermodynamic gaps in the bands which are non-topological in nature. This can be considered as remnants of the finite gaps which occur in even when simple tight binding model is implemented on Sierpinski gasket (see Fig. 4.1).

Although the states close to $E = 0$ resemble "edge" states, the Bott index, a topological invariant adapted to these systems [22], is found to be zero—signaling a trivial phase (Fig. 4.4).

4.3 Transport and Chiral States

What is nature of transport in this gapless phase? Is this a usual metal? To investigate this we employ two methods. We construct an initial state on a corner of the Sierpinski gasket and project it onto the occupied states between the energy(E) interval $\{-(2 + M) < E < E_F\}$. These primarily comprise of 'edge' states. The time evolution of this state is shown in Fig. 4.5a. It is clear that the wavepacket moves only in a

Fig. 4.4 Thermodynamic gaps in bulk states: The spectrum for the Sierpinski gasket when the topological Hamiltonian is implemented on it. Results for different generations are plotted for $M = -1$. These gaps are however non-topological (Bott index is *zero*)

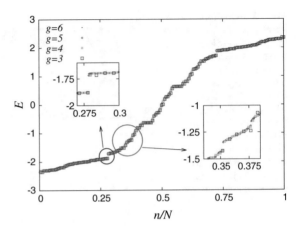

particular direction and is confined to the "edge". Therefore the metallic phase is a chiral metal rather than a trivial metal! It is also interesting to note that at the point of contact can allow for splitting of the wavepacket as is seen in last image. To further investigate the transport properties, we connect the gasket to leads and calculate the two terminal conductance using the Non-Equilibrium Green's Function method (NEGF) [23]. The results are shown in Fig. 4.5b. One finds that the dimensionless two terminal conductance ($\frac{Gh}{e^2}$) is close unity near the band center, but, not exactly. One can also notice that the conductance has sharp dips as a function of energy. These coincide with the edge states which reside on the "inner edges". Although these are "edge" states, they do not contribute to two terminal transport.

This system, where the bulk and the boundary, are not clearly defined has the following curious features: its spatial dimension is 1.58. The periodic case, where each site is equivalently coordinated, is topologically trivial (as Bott index is zero) and in fact is a gapless metal at half filling. However, this is not a trivial metal. The states are chiral, and conduct close to the quantum of conductance when the states lie at the outer edge. Other states also resemble edge states—where they reside at inner edges. The edge spectrum is in fact itself self-similar. It will be interesting to investigate how the concept of Berry curvature can be extended to such a system.

What is the physics behind all this? While the story may be far from complete, it is useful to conjecture a scenario why this can happen. Fractal dimensional objects can come in two varieties—first, where more structure is built on a lower, yet integral dimensional object, such as a Koch curve. Another way is to consider a higher dimensional object and delete parts of it recursively. The first procedure is not expected to change topological features of an object—since no extra boundaries are created. In the latter case, the scooping out can only be accompanied by creation of boundaries which implies creation of further robust edge states at every generation. Existence of these 'edge' states at all scales will construct chiral states. Sierpinski gasket is an example of the second kind of fractal. In fact the chirality of a wavepacket created in one of the inner edges is opposite to that of the outer edge, allowing for

Fig. 4.5 Transport: a A
wave packet created at the
top-right edge of the gasket
is projected on the occupied
states within the energy
range $(-(2 + M) < E <
E_F = 0)$. The sequential
panels show the evolution of
such a packet. It can be seen
that it moves exclusively on
the edge of the system with a
particular chirality. Here
$g = 4$, $M = -1$. **b** Two
terminal conductance
through the Sierpinski gasket
when it is in the gapless
regime $(M = -1)$

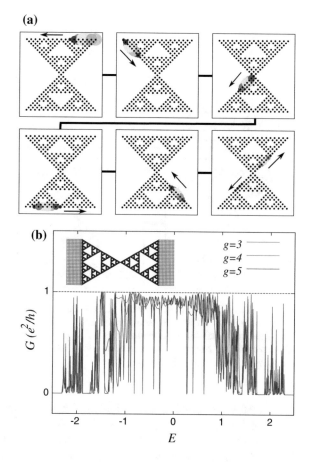

some semblance of this physics to a scooped quantum Hall system. This is shown in
Fig. 4.6.

Of all kinds of "edge" states, the outermost one is delocalized on a larger area
and therefore is closest to the band center. These states will therefore conduct as was
seen in the NEGF calculations. The only way to gap out these outermost states is to
patch Sierpinski triangles in ways which do not keep the same coordination number
intact. As an example we discuss two such systems in the next section.

4.4 Sierpinski Carpet and Torus; Bott Index

We now consider two more fractal systems, but where every site is not equivalently
coordinated. We set up the same Hamiltonian as mentioned in Eq. (4.1) and calculate
the topological index as a function of the parameter M. For the first system, we

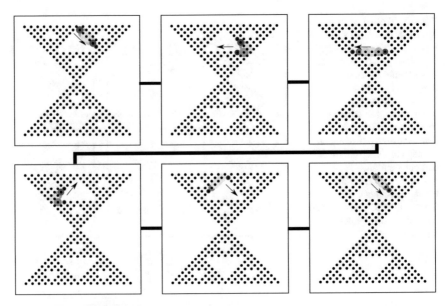

Fig. 4.6 Transport in an inner edge: a A wave packet created at the top inner edge of the gasket is projected on the occupied states within the energy range $(-(2+M) < E < E_F = 0)$. The sequential panels show the evolution of such a packet. It can be seen that it moves exclusively on the edge of the system with a particular chirality. Here $g = 4$, $M = -1$. Interestingly the chirality is opposite to the one seen in Fig. 4.5

combine four Sierpinski triangles into the form as shown in Fig. 4.7. One can notice here that the sites belonging to the boundary of the triangles have a larger coordination number. The variation of the Bott index is also shown at half filling. One finds that in this system, under periodic boundary conditions, edge states does not appear at the outermost edge and the Bott index is nontrivial in a regime of M. This coincides with existence of edge states on the outermost edge in an open system. The edge states corresponding to the inner edges continue to remain both for open and closed systems.

The second system which we analyze is the Sierpinski carpet. The system is shown in Fig. 4.8. Here 1/8th of the carpet is scooped out recursively. The Hausdorff dimension for this system is 1.8928. We again set up the same topological Hamiltonian on this system, and find the Bott index as a function of the parameter M. Here again every site is not equivalently coordinated. Another crucial difference between Sierpinski carpet and gasket is the concept of ramification. Ramification counts the number of distinct bonds which need to be deleted to break the fractal into macroscopic objects. For Sierpinski gasket this number is 4. For the Sierpinski carpet this number is infinity. It will be interesting to see, how ramification effects realization of topological phases on fractals.

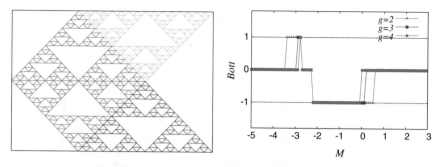

Fig. 4.7 Sierpinski Torus: (Left) Four Sierpinski triangles can be combined to make a structure which can be made into a torus. (Right) Bott index, when calculated at half filling, is calculated as a function of M. One finds a nontrivial regime

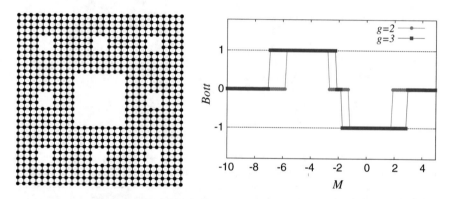

Fig. 4.8 Sierpinski Carpet: (Left) Sierpinski carpet is a recursively scooped out square lattice. The picture shown is for a third generation system. (Right) Under periodic boundary conditions, and Bott index can be calculated as a function of M. The system shows a nontrivial topological regime

4.5 Sierpinski Tetrahedron

We now illustrate another example of a fractal system which shows intriguing features and avoid interpretation in terms of "usual" topological systems. We construct a three dimensional version of the Sierpinski gasket where one-fourth of a tetrahedron is recursively scooped out, called the Sierpinski tetrahedron. Such a system, though embedded in a three dimensional space, has a Hausdorff dimension of 2. The top vertices of this system can be identified with the bottom three points—again producing a "periodic" system in the sense that every site in this system has the same coordination number 6. While one might naively expect that this system should be a straight forward generalization of the previous example, this is rather subtly different. The edges in the Sierpinski gasket can be considered "one-dimensional" as every inner triangle has a one-dimensional perimeter. The Sierpinski tetrahedron has infinite

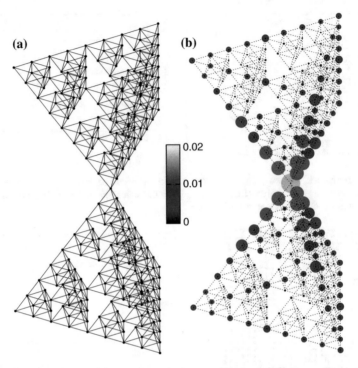

Fig. 4.9 Sierpinski Tetrahedrons: a Recursively scooped out tetrahedrons can be combined to form a Sierpinski tetrahedron. While the system is embedded in three dimensions, its Hausdorff dimension is 2. The figure shown is for $g = 4$. **b** Setting up of a topological Hamiltonian results in formation of surface states which look like connections made of various "corner" states

number of surfaces, however, each of these surfaces are itself fractals—Sierpinski gaskets. Setting up a four-orbital model of a topological insulator in this system (see Eq. (3.18)), one finds that in a regime of parameters, close to $E = 0$ one can have "surface" state which looks like a network of corner states (see Fig. 4.9). With increasing generation, one also finds that this system is a metal. This same model, in the same parameter regime, is known to produce surface states in a cubic lattice [24] and even on an amorphous system as shown in the last chapter.

4.6 Perspective

In conclusion, in this chapter, we have investigated fractals with the aim of seeking "topological" phases in them. Our investigations in the Sierpinski gasket unequivocally show that the system has a metallic phase which is chiral in nature. The regime of this coincides with a conventional lattice showing topologically insulating phases. The behavior can be qualitatively understood as a result of proliferation of edges due to the geometry of the structure. However a periodic system in this context, intuitively discussed here as a system where every site is equally coordinated, cannot be made topologically insulating. The only scope to obtain such a phase is to break this equivalence of sites. We also investigated a "three" dimensional system whose surfaces are themselves fractals. These investigations presses the question as to where should such phases be placed in Kitaev's table of topological classification. It also urges renewed investigations on the bulk-edge correspondence and its implications on systems such as fractals.

References

1. Hasan MZ, Kane CL (2010) Colloquium: topological insulators. Rev Mod Phys 82:3045–3067
2. Qi X-L, Zhang S-C (2011) Topological insulators and superconductors. Rev Mod Phys 83:1057–1110
3. Chiu CK, Teo JCY, Schnyder AP, Ryu S (2016) Classification of topological quantum matter with symmetries. Rev Mod Phys 88:035005
4. Ludwig AWW (2016) Topological phases: classification of topological insulators and superconductors of non-interacting fermions, and beyond. Phys Scr 2016(T168):014001. http://arxiv.org/abs/1512.08882
5. Thouless DJ, Kohmoto M, Nightingale MP, den Nijs M (1982) Quantized hall conductance in a two-dimensional periodic potential. Phys Rev Lett 49:405–408
6. Domany E, Alexander S, Bensimon D, Kadanoff LP (1983) Solutions to the Schrödinger equation on some fractal lattices. Phys Rev B 28:3110–3123
7. Mandelbrot BB, Pignoni R (1983) The fractal geometry of nature. WH Freeman, New York
8. Gefen Y, Mandelbrot BB, Aharony A (1980) Critical phenomena on fractal lattices. Phys Rev Lett 45:855–858
9. Gefen Y, Aharony A, Mandelbrot BB, Kirkpatrick S (1981) Solvable fractal family, and its possible relation to the backbone at percolation. Phys Rev Lett 47:1771–1774
10. Rammal R, Toulouse G (1982) Spectrum of the Schrödinger equation on a self-similar structure. Phys Rev Lett 49:1194–1197
11. Alexander S, Orbach R (1982) Density of states on fractals:fractons. Journal de Physique Lettres 43(17):625–631
12. van Veen E, Yuan S, Katsnelson MI, Polini M, Tomadin A (2016) Quantum transport in Sierpinski carpets. Phys Rev B 93:115428
13. Alexander S (1984) Some properties of the spectrum of the Sierpinski gasket in a magnetic field. Phys Rev B 29:5504–5508
14. Fukushima M, Shima T (1992) On a spectral analysis for the Sierpinski gasket. Potential Anal 1(1):1–35
15. Wang XR (1995) Localization in fractal spaces: exact results on the Sierpinski gasket. Phys Rev B 51:9310–9313
16. Chakrabarti A (1996) Exact results for infinite and finite Sierpinski gasket fractals: extended electron states and transmission properties. J Phys Condens Matter 8(50):10951
17. Pal B, Chakrabarti A (2012) Staggered and extreme localization of electron states in fractal space. Phys Rev B 85:214203
18. Gordon JM, Goldman AM, Maps J, Costello D, Tiberio R, Whitehead B (1986) Superconducting-normal phase boundary of a fractal network in a magnetic field. Phys Rev Lett 56:2280–2283
19. Shang J, Wang Y, Chen M, Dai J, Zhou X, Kuttner J, Hilt G, Shao X, Gottfried JM, Wu K (2015) Assembling molecular Sierpiński triangle fractals. Nat Chem 7(5):389–393

20. Song ZG, Zhang YY, Li SS (2014) The topological insulator in a fractal space. Appl Phys Lett 104(23):1–5
21. Bernevig BA, Hughes TL (2013) Topological insulators and topological superconductors. Princeton University Press, Princeton
22. Loring TA, Hastings MB (2010) Disordered topological insulators via C*-algebras. Eur Phys Lett 92(6):67004
23. Datta S (1997) Electronic transport in mesoscopic systems. Cambridge University Press, Cambridge
24. Fradkin E (2013) Field theories of condensed matter physics. Cambridge University Press, Cambridge

Chapter 5
Killing the Hofstadter Butterfly

In the last chapter we looked at fractals, where the spatial dimension is itself not an integer. We looked at the construction of a topological model on such a system and found that one finds a fractal spectrum where the eigenenergies are self similar. Here we construct a system which is otherwise translationally invariant, but has a fractal spectrum. We then investigate—what happens to this system if we remove bonds randomly?

Understanding the role of disorder on electronic conduction has been a central theme in all of condensed matter physics [1–5] as we had discussed in Chap. 1. Apart form being fundamentally interesting from a theoretical perspective, these problems hold immense significance as they directly bring out (or hide) novel physics in various experimental systems [6]. One of the essential probes in condensed matter is the magnetic field. Effects of which on a $2D$ electron gas leads to integer and fractional Hall effect [7, 8]. The same phenomena on an idealized square lattice leads to the Hofstadter model [9]. Interestingly this physics has now been realized both in cold-atomic systems [10, 11] and material systems [12, 13]. These systems possess nontrivial topology (see Sect. 1.3) and robust signatures in transport [14, 15]. Not surprisingly, the effect of disorder on quantum Hall physics has received its due attention [16–19]. For the continuum model—this question can be posed in two ways—how does the conductance change when, while keeping the magnetic field same, the disorder is increased or; keeping the disorder same, the magnetic field is reduced. The evolution of the Landau levels, in a 2D electron gas, and in the lattice setting with a weakening magnetic field has been a matter of debate [20, 21]. For a recent review refer to [22]. It was earlier suggested that to be consistent with the scaling hypothesis [23], the Landau levels will float up to higher energies with decreasing magnetic field or increasing disorder [24–26]. However some numerical calculations have hinted otherwise and have instead suggested that the system undergoes a Chern insulator to normal insulator transition as a function of the strength of the disorder [27]. A two parameter scaling theory has been suggested to understand this transition and a phase diagram was also proposed [28–30].

© Springer Nature Switzerland AG 2019
A. Agarwala, *Excursions in Ill-Condensed Quantum Matter*,
Springer Theses, https://doi.org/10.1007/978-3-030-21511-8_5

However disorder comes in various varieties—Anderson disorder [1] is the most famous of them all. In this, random onsite potentials are added to each site of the lattice. The other more stronger kind of disorders are the percolation disorders. They come in two varieties—site and bond. In the former, one randomly removes the sites from the lattice, in the latter, bonds. Till date, quantum site and bond percolation in 2D even in absence of a magnetic field is poorly understood and highly debated— the central question being—whether the physics here is different from Anderson disorder [31, 32]. In fact delocalization-localization transition has been predicted in 2D for site-dilution on square lattices [33–36]. In this work we limit ourselves to the discussion on bond percolation. If we define p_b as the probability of a link being present between two neighbouring sites, then in classical bond percolation, percolation threshold occurs at $p_c = 1, 0.5$ and 0.2488 for hyper-cubic lattices in dimensions(D) $= 1, 2$ and 3 respectively [37, 38]. This threshold signifies the point below which there exists no geometrical connecting path between two sides of a lattice. One expects quantum bond percolation threshold p_q to be $> p_c$ since interference effects will tend to further localize the system even if classically a path may exist. Notice that unlike Anderson disorder here exists a natural bound on p_q due to presence of p_c. While finite size scaling analysis shows an existence of a percolation threshold p_q in 3D [39], results for 2D are still not settled [32]. Some of the previous works have predicted nonzero conductance for $p > p_c$ while others have predicted that all states get localized even for infinitesimal disorder [39–43]. A study of transport in bond percolating system and its comparison with classical Drude theory expectations have also been performed [44]. Recently, bond (and site) percolation on a honeycomb lattice has received major attention in order to understand the nature of divergence of density of states at $E = 0$ [45–48].

As far as the effect of magnetic field is concerned, most of the studies above [27, 29, 30] has been performed for diagonal Anderson disorder. A study of banded off-diagonal disorder was performed in [49]. However the role of percolation disorder on the Hofstadter model has been little investigated. A periodic dependence of p_q was found as a function of magnetic flux in 3D while that in 2D was also conjectured [50]. Since for bond percolation disorder the exact value of p_q itself is an open question, it is of particular interest to find if there exists a metal insulator transition before we cross the classical percolation threshold in presence of a magnetic field.

In this chapter, our motivation is two fold. The first part involves understanding the effect of bond percolation disorder on the Hofstadter butterfly pattern as a function of p_b. We study the model in both high and low concentration of bond dilutions. We find that even at high amount of bond dilution, we have butterfly-like patterns present in the system. We also look at the effect of bond dilution on band gaps and wave functions of the system. We provide understanding of the key features of our results from analyzing small clusters and finite size rings. This provides some physical reasoning behind the results and also contrast them from the case of Anderson disorder. The second part involves calculation of the transport quantities (σ_{xy}), where a numerical study based on calculation of Chern numbers is performed using coupling matrix approach [51]. We study the effect of bond percolation disorder on Hofstadter bands and show that there indeed is a metal insulator transition with decreasing p_b for

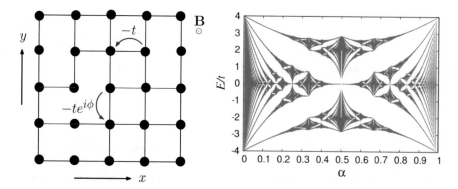

Fig. 5.1 The model and the pattern: (Left) A schematic of a square lattice with some of the bonds removed. p_b is the probability that a bond is present. Therefore, at $p_b = 1$ we have an ideal square lattice. t is the hopping amplitude which is set to 1. ϕ is the Peierls phase which includes the effect of the magnetic field **B**. (Right) The plot of the energy dispersion as a function of the magnetic flux α. This is the Hofstadter butterfly as was originally reported by Hofstadter [9]

$p_b > p_c$. We also find that the Chern bands close to the band edges are more stable to disorder than the ones close to band center, which means that it takes higher disorder strength for achieving metal to insulator transition at the edge of the band than at the center. This result in the low-disorder limit is consistent with the findings for the Anderson disorder case [27].

We now present the plan of this chapter. In the next section (Sect. 5.1) we provide a brief review of the Hofstadter model and present some of the results for finite rings in presence of magnetic field. These will be used later in our study. In the same section we also introduce the percolation problem. Section 5.2 contains our results and related discussions on the effect of bond percolation disorder on the Hofstadter butterfly. Here we also discuss the effect of bond dilution on band gaps and on wave functions using inverse participation ratios (IPRs). Section 5.3 contains the essential details about the coupling matrix approach to calculate the Chern number in the presence of disorder and the corresponding results and discussions. In Sect. 5.4 we summarize our results and speculate some future directions.

5.1 Formulation and Prelude

5.1.1 Hamiltonian

The Hamiltonian of our interest is

$$\mathcal{H} = \sum_{\langle i,j \rangle} -t e^{i\phi_{ij}} c_i^\dagger c_j + h.c. \tag{5.1}$$

where $c_i^\dagger$, c_j are the creation and annihilation operators for the electrons at site i and j respectively (see Fig. 5.1). The $\langle i, j \rangle$ signifies that the sum is over the nearest neighbors on a square lattice. ϕ_{ij} is the Peierls phase which takes into account the effect of a perpendicular magnetic field B on the lattice and is given by

$$\phi_{ij} = \frac{e}{\hbar} \int_{r_j}^{r_i} A \cdot dr, \tag{5.2}$$

where A is the corresponding vector potential. t is the hopping integral and is set to 1. $r_{i(j)}$ denotes the position coordinates of site $i(j)$. We work in Landau gauge where $A = (0, Bx, 0)$. This conveniently allows for complex phases only in the hoppings in the vertical direction. The flux per plaquette is given by α in units of h/e. We will ignore the spin of the fermions.

5.1.2 Hofstadter Butterfly

In the gauge we are using, the system has a translational symmetry in y direction— therefore k_y is a good quantum number. For a generic α, the system does not have translational symmetry in x direction. However, when $\alpha = p/q$, where p, q are integers, the problem can be mapped to a reduced Brillouin zone. The eigenvalues can be plotted as a function of α and this leads to the famous Hofstadter butterfly pattern. This self-similar, fractal pattern was first obtained by Hofstadter [9], and is reproduced in Fig. 5.1.

5.1.3 Polygon in a Magnetic Field

Next, let us consider a N sided polygon in a magnetic field. The eigenvalues indexed by M are given by

$$E_N(M, \alpha_p) = -2t \cos\left(\frac{2\pi}{N}(M + \alpha_p)\right) \tag{5.3}$$

where $M = \{0, 1, \ldots N - 1\}$ and α_p is the flux going through the polygon [52]. Note that α_p is different from the flux per unit plaquette α as introduced in the previous subsection. Figure 5.2 shows the dispersion for few representative finite size rings in presence of uniform magnetic field. As can be seen from Fig. 5.2c, d, both the polygons have 8 sides, but the total flux inside the loops are different. While (c) has $\alpha_p = 3\alpha$, the latter (d) has $\alpha_p = 4\alpha$. These lead to different dispersions ((g)–(h)). These as we will see later will be useful in understanding the results in presence of percolation later.

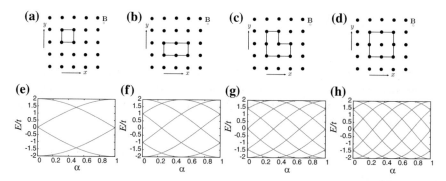

Fig. 5.2 Rings in magnetic field: **a** A connected square with a unit flux α per plaquette has a dispersion shown in **e**. **b** A polygon with 6 sides, has two unit squares inside, this corresponds to $\alpha_p = 2\alpha$ in Eq. (5.3) and has a dispersion in **f**. A 8-sided polygon can have $\alpha_p = 3\alpha$ (shown in **c**) and $\alpha_p = 4\alpha$ as shown in **d**. The corresponding dispersions are shown in **g** and **h** respectively

5.1.4 Disorder and Percolation

Next we define what we precisely mean by percolating the lattice. p_b is defined as the probability of a link being present between two neighboring sites. This implies that at $p_b = 1$ we have an ideal square lattice. For any value of $p_b < 1$ some of the bonds are removed from the lattice (see Fig. 5.1). For a square lattice there is a classical percolation threshold at $p_b^c \equiv p_c = \frac{1}{2}$. At any value of $p_b < p_c$ there does not exist a classical geometrical path connecting the two sides of the square lattice [32]. Percolation transitions have their own universality classes and distinct critical exponents [38]. Percolation is therefore, a special kind of disorder. Even in quantum transport, note that each bond removal is of the energy scale t which is of the same order as the bandwidth. However density of bonds removed, quantified as $(1 - p_b)$, is considered as the tuning parameter of disorder strength.

Another kind of percolation problem is the site percolation problem. Here sites are randomly removed from a lattice. We state that although, both bond and site percolation problems retain the sublattice symmetry, the bond problem has some 'nicer' features than the site percolation. Once a site is removed from a lattice, it effectively reduces the Hilbert space of the problem. Given a imbalance between the number of sites belonging to the two sublattices, one finds *zero* energy modes in the system, which need to be removed 'by-hand' to keep track of nontrivial zero modes [45, 53]. On the contrary, removing bonds on the lattice keeps the dimension of Hilbert space same and only modifies the connectivity between the sites.

As was mentioned in the introductory section, the most well studied disorder problem is the Anderson disorder [1]. Here onsite potentials to each site is chosen randomly from a distribution (mostly 'box') between $[-\frac{W}{t}, \frac{W}{t}]$. Thus W is the parameter characterizing the strength of disorder. A review of numerical results on this can be found in [54].

5.2 Killing the Butterfly

In Fig. 5.3 the evolution of the Hofstadter butterfly as function of p_b for some representative values of p_b is shown. While $1 - p_b$ can be considered as the 'strength' of disorder (like W in Anderson disorder case) we will see that both these disorders are quite different in high disorder limit. Let us first look at the case when p_b is very small and $p_b \ll p_c$ (high disorder limit).

5.2.1 $p_b \ll p_c$

In this limit, the system is below the classical percolation threshold and therefore the lattice has already geometrically broken up into disconnected fragments. As can be seen from Fig. 5.3j, l one finds that there are many bands which do not disperse with α. This can be understood from the fact that most of these structures do not have closed loops which have any magnetic flux passing through.

We also see that the energies cluster around specific values. To understand these, consider a single site $(j = 1)$ connected to N $(j = 2, \ldots, N + 1)$ other sites with an equal hopping strength $-t$ and no other site is connected to any other. Let the eigenvalues be ε_i, where $(i \in 1, \ldots N + 1)$. For a generic N, one finds only two nonzero eigenvalues given by $\pm\sqrt{N}t$. The corresponding eigenvectors are $\frac{1}{\sqrt{2}}(\mp 1, \underbrace{\frac{1}{\sqrt{N}}, \ldots, \frac{1}{\sqrt{N}}}_{N})^T$. The other eigenvectors corresponding to *zero* eigenvalues are of the form $\frac{1}{\sqrt{2}}(0, 0, .., \underbrace{1}_{i}, .., \underbrace{-1}_{j}, \ldots 0)^T$, where i, j denotes the site index and take the values $\in (2, \ldots, N + 1)$ and therefore has $N - 1$ solutions. Since the maximum coordination number for a square lattice problem is 4, the corresponding nonzero eigenvalues are $\pm t(N = 1)$, $\pm\sqrt{2}t(N = 2)$, $\pm\sqrt{3}t(N = 3)$ and $\pm 2t(N = 4)$. Note that all of these structure have no loops and therefore, the eigenenergies will not change with α. The probability of these structures appearing are $\propto p_b^N$ [38]. This therefore also implies that in this limit we have segregation of eigenvalues at some set of discrete energies and DOS peaks only at these specific energy eigenvalues.

Note that this limit of the Hofstadter model in presence of bond percolation is absolutely distinct from Anderson disorder. The connectivity of each lattice point to the other is not changed in the case of Anderson disorder, and therefore at no value of W do we expect nondispersing eigenvalues (with α). Similarly, increase in W will never lead the eigenvalues to segregate at select eigenvalues. On the other hand, in bond percolation, at $p_b = 0$ the DOS will show a δ function peak at $E = 0$.

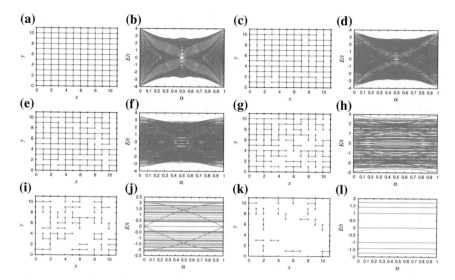

Fig. 5.3 **Killing the pattern**: The representative lattices of 12×12 size at different values of p_b and their energy dispersion as a function of the magnetic flux α. The lattices shown in **a** belongs to $p_b = 1$. For **c** $p_b = 0.9$, **e** $p_b = 0.75$, **g** $p_b = 0.50$, **i** $p_b = 0.25$ and **k** $p_b = 0.1$. The corresponding dispersion as a function of α is shown in **b**, **d**, **f**, **h**, **j** and **l** respectively

5.2.2 $p_b < p_c$

As p_b is slightly increased, as can be seen in Fig. 5.3j, h, dispersing bands appear. While the complete lattice still does not have a spanning cluster, what is clear is that we have states in the system which disperse with magnetic flux α. These are due to small clusters which contain closed loops. Take for example the representative plot shown in Fig. 5.3j and compare the dispersing curve with the Fig. 5.2e. As can be seen they are exactly the same. Therefore, the low p_b "Hofstadter butterfly" will be dominated only by these finite size small loops, as shown in Fig. 5.2. Note that all these states cannot contribute to transport since they reside only on small clusters.

5.2.3 $p_b = 1$

We now discuss the other limit i. e. the clean system. Clearly, even the finite Hofstadter butterfly as shown in Fig. 5.3b has some semblance to the infinite Hofstadter butterfly as shown in Fig. 5.1, increasing lattice size makes this similarity more and more apparent [52]. However, even the finite lattice system has some interesting gap structure at $E = 0$ which we now discuss (see Fig. 5.4). Any finite size square lattice of dimensions $(L \times L)$ shows a number of bands dispersing linearly from $E = 0$. This number and the slope increases in an interesting fashion, which can be guessed from

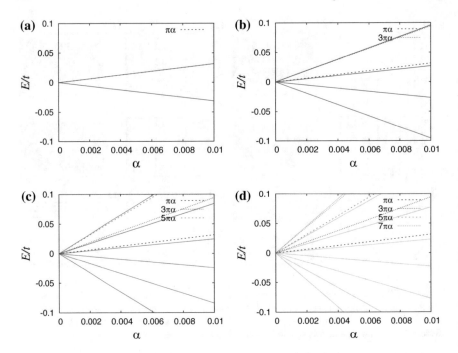

Fig. 5.4 Gapless states: Band dispersion for square lattices with open periodic condition for lattices of size $L \times L$ for **a** 2×2 **b** 4×4 **c** 6×6 and **d** 8×8 for α and E close to *zero*. The slope of dispersion approximately follows the slope of $\pi M (L - 1)\alpha$

our discussions on the finite size polygons in magnetic field. Note that a $L \times L$ square ring has $(L - 1)^2 \alpha$ flux passing through it. Substituting $N = 4L - 4$, $M = N/4$ in Eq. (5.3), we find the low energy dispersion of the form

$$= -2t \cos \left(\frac{2\pi}{N} (2(L - 1) + (L - 1)^2 \alpha) \right) \tag{5.4}$$

$$\approx (L - 1)\pi \alpha. \tag{5.5}$$

Now a square lattice of $L \times L$ can contain states on concentric square rings of dimensions $2, 4, \ldots, L$, where the states on this rings have small α dispersion as $\pi \alpha, 3\pi \alpha \ldots (L - 1)\pi \alpha$ near $E = 0$. This can be clearly seen from Fig. 5.4. As is expected the states indeed lie predominantly on the concentric rings.

5.2.4 $p_b \gg p_c$

We now look at the effect of the low bond disorder on the finite size Hofstadter butterfly. We focus on the band gap structure at $E = 0$. We see from Fig. 5.5, a pure

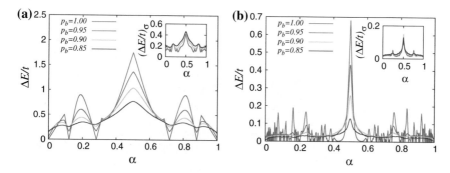

Fig. 5.5 The gaps: The band gap at $E = 0$ for four values of $p_b = 1.0, 0.95, 0.90$ and 0.85 as a function of α for a **a** 4×4 lattice and **b** 12×12 lattice. For the later three values of p_b, averages are being plotted over 400 configurations. The *blue* line is for the pure system, and has many gapless points and large band gaps including one at $\alpha = 0.5$. Increasing the disorder, opens up the gap at gapless points and reduces the magnitude of the larger gaps from the pure system. The lower the value of p_b, the effect is larger. In the inset, the variance ($\equiv (\Delta E/t)_\sigma$) for the later three values of p_b are plotted. It shows the variance $\Delta E/t$ increase with decrease in p_b. All these are consistent with our understanding that generic weak disorder spreads out the *DOS* and open up gaps at the gapless points

4×4 and 12×12 ((a) and (b)) lattice size has a set of gapless points and large band gaps at some other values of α. Increasing disorder, opens up gaps at the gapless points and reduces otherwise large band gaps. This, in some sense, is the usual effect of any disorder i. e. spreading of the *DOS*. It is also reasonable to see that this effect increases with increasing disorder. This is more clear from the inset in the Fig. 5.5 where the variance of the gap is plotted. We also note that the amount of gap opened up at $\alpha = 0$ is much smaller than other gapless points. To further understand the effect of bond percolation disorder, we study the Inverse Participation Ratio (IPR) of the different wave functions. IPR for a unit normalized wave function $|\psi\rangle$ expandable in site basis as $|\psi\rangle = \sum_i \psi_i |i\rangle$ is given by

$$IPR = \sum_i |\psi_i|^4. \tag{5.6}$$

This value estimates the spread of a wave function in real space. For a delocalized wave function spread uniformly over area A, IPR $\propto 1/A$, and will decrease with increasing area. If a wave function is localized over some few sites, then IPR $\propto 0.1$–1 and doesn't change significantly with increasing size of the system. This diagnostic therefore provides a scope to demarcate localized and delocalized states. To estimate the effect on IPR, we consider 400 configurations of the lattice at a given p_b and calculate the state index averaged IPR and the corresponding energy. The results are shown in Fig. 5.6. As the p_b is reduced, IPR at certain values of E becomes very large. Interestingly the values are at $\pm t$, $\pm\sqrt{2}t$ and $\pm\sqrt{3}t$. These correspond to clusters of small sites mentioned before. The corresponding IPRs for these is 1,1/2 and 1/3.

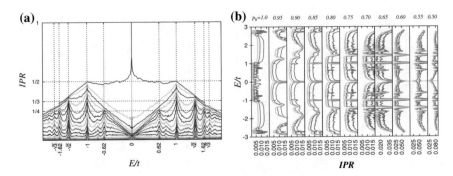

Fig. 5.6 Variation of IPR as a function of bond probability: **a** IPR is shown for a 30×30 lattice for p_b starting from 1.0 (bottom-most) to 0.05(topmost) in intervals of 0.05. One notices appearance of peaks at specific values of E/t which goes away with reducing p_b. (see text) **b** IPR for system sizes 18×18 (blue), 24×24 (orange) and 30×30 (yellow) compared to each other at $\alpha = 1/4$ when averaged over 400 configurations. At $p_b = 1.0$ the IPR is quite small ($\sim 5 \times 10^{-3}$) and reduces with increasing system size. Decreasing p_b one notices a strong peak appears at $E = 0$, implying appearance of localized states. However, one notices that decreasing p_b more high IPR peaks start to appear at distinct energy values. Also the average IPR of the system increases for a 30×30 lattice and overlaps with the case of 24×24, implying overall localization in the spectrum. This feature becomes prominent for $p_b \lesssim 0.65$. Error bars are not shown for clarity of figure

As p_b is decreased further some of these peaks vanish, and now only the central peak remains. Also peaks at other values correspond to the solutions for a open tight binding chain. For a n-site chain the dispersion is given by $-2t\cos k$, where $k \in \frac{m\pi}{n+1}$, where $m \in (1, 2 \ldots n)$. For example, a 4-site open chain has eigenvalues at $\pm 2\cos(\frac{\pi}{5})(\sim \pm 0.62)$ and $\pm 2\cos(\frac{2\pi}{5})(\sim \pm 1.62)$ which can also be clearly seen in Fig. 5.6a.

In Fig. 5.6b we look at relatively smaller values of p_b and look at the effect of increasing the system size. The variation of IPR signifies whether the system is comprised of localized or delocalized states. For example at $p_b = 1.0$ one finds that IPR is quite low ($\sim 5 \times 10^{-3}$) and reduces with increasing system size. Decreasing p_b one notices a strong peak appears at $E = 0$, implying appearance of localized states. However, one notices that decreasing by p_b high IPR peaks start to appear at distinct energy values as discussed in detail above. However, interestingly the average IPR of the system increases, and the value for 30×30 starts to overlap with 24×24, implying localization overall in the full spectrum. This feature becomes quite prominent at $p_b \lesssim 0.65$. Note that the average IPR is about 0.025 at $p_b = 0.60$ signaling that the wave function resides only on a average of 40 sites, in otherwise a lattice of 900 sites. This signals the wave functions have got localized much before the classical percolation threshold is reached. This will be investigated more clearly through calculations of the transport in the next section.

5.3 Effect on Chern Numbers and Transport

Hofstadter Model, apart from structure of the eigenspectrum, also hosts interesting structure of the topological invariants [14, 15, 55]. It will be interesting to understand the effect of disorder on such topological invariants. This therefore requires calculation of Chern numbers. Note that in presence of disorder, the system no longer contains translational symmetry and therefore a momentum integral over the Brillouin zone will not suffice to calculate the Chern number. We therefore calculate the Chern numbers using the method outlined in [51]. This essential numerical technique is motivated from the fact that Chern number can also be calculated from an integral over the twisted boundary conditions. For completeness we include briefly some of the definitions and a brief discussion about the method following [51].

For a $2D$ lattice comprising of $N = L \times L$ unit cells, the single particle wave functions can satisfy the following boundary conditions given by $\phi_\theta(x + L, y) = e^{i\theta_x}\phi_\theta(x, y)$ and $\phi_\theta(x, y + L) = e^{i\theta_y}\phi_\theta(x, y)$, where $\theta = (\theta_x, \theta_y)$ such that $0 \leq \theta_x, \theta_y \leq 2\pi$. For a given filling, we can have M states occupied. Let the many body wave function of these M states be written as Ψ_θ. Then the Chern number of the ground state is given by

$$C = \frac{1}{2\pi i} \int_{T_\theta} d\theta \langle \nabla_\theta \Psi_\theta | \times | \nabla_\theta \Psi_\theta \rangle \qquad (5.7)$$

where T_θ denotes the allowed (θ_x, θ_y) values [56]. Note that since in defining Ψ we have taken into consideration all the filled states, C here is the sum of the Chern numbers of individual bands below the chemical potential. Hence the quantity evaluated can be interpreted as σ_{xy} in units of e^2/h.

The calculation of σ_{xy} can be directly done using Lehmann representation of the Kubo formula [57]. However finding the Chern number of each band requires numerical diagonalization of a system many number of times [27]. The recently developed coupling matrix approach allows for a much simpler and numerically inexpensive method [51]. The idea is to convert the integral over T_θ into an integral over a path in momentum space. This integral then can be solved as a product of matrices whose components are determined by the inner product of some wave functions which were determined only by the system under periodic boundary conditions. The essential simplifying step is to do away with the necessity of diagonalizing the system at different values of boundary conditions. This approach can also take into consideration the effect of real space disorder in a natural way.

Now we present the results of our calculations. In Fig. 5.7 we show the variation of σ_{xy} with bond occupation probability p_b for $p/q = 1/4$. The filling is kept constant at 1/4 and for each configuration chemical potential is self consistently evaluated. The lattice size is systematically increased from 12×12 to 24×24 in difference of 4 sites per side. We find that increasing lattice size makes the transition sharper and clearly the conductance goes to zero much before p_c which occurs at $p_b = 0.5$. The transition seems to occur close to $p_b \approx 0.65$, which was also tentatively the value seen from IPR results in previous section. In Fig. 5.7 we also show the variation of

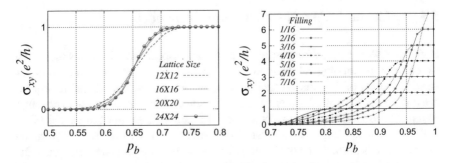

Fig. 5.7 The transition: (Left) Plot of σ_{xy} with bond occupation probability p_b at $p/q = 1/4$ for different lattice sizes. The filling is kept constant at 1/4 and the lattice size is increased from 12×12 to 24×24. The red line with moon points is for the lattice size of 24×24. With increasing lattice size we find the transition becoming sharper around $p_b \approx 0.65$. The *standard error* of the mean is of the order of the point size or lower and hence has not been shown above. (Right) σ_{xy} with bond occupation probability p_b for $p/q = 1/16$ for different fillings. The lattice size is 32×32. In the clean limit ($p_b = 1$) σ_{xy} for fillings $n/16$ is $n\frac{e^2}{h}$ where ($n \in 1, \ldots, 7$). With increasing filling one moves from the bottom of the full spectrum to the center. The Chern insulator plateaus are less stable to bond disorder as one moves closer to band center. The results for both the plots are averaged over 400 disorder configurations

σ_{xy} with bond occupation probability p_b for $p/q = 1/16$ for different fillings. We find that with increasing filling the Chern insulator plateau remains stable only for lower strength of bond disorder. Note that with increasing filling from 1/16 to 7/16 we moved from bottom of the spectrum to band center. This resembles what was found for the Anderson disorder in earlier studies [27].

To understand the underlying mechanism for this, one first realizes that the physics of the Hofstadter problem is subtly different from that of the continuum 2D model. Unlike the continuum, σ_{xy} can be negative in the lattice setting [55]. This is because the bands here may carry negative Chern numbers. For an even q a negative Chern number band of Chern number $-2(q - 1)$ lies at the band center. With increasing onsite disorder is has been argued that this central band mixes with the other bands hence explaining why the bands close to band center are the first ones to show transition to normal insulator [27]. We infer that the similar considerations indeed apply in this limit of bond percolation problem.

Also there is a lot of interest in understanding the physics in the limit of *zero* magnetic field. We expect that with weakening magnetic field, the amount of bond percolation disorder required for Anderson insulating transition will decrease i. e. p_q will slowly approach a larger value. This can be expected from Fig. 5.7. The magnetic flux is kept high ($p/q = 1/4$) in the first case and the transition occurs at $p_b \approx 0.65$. While in the second case the magnetic flux is kept low ($p/q = 1/16$), and the transitions for all values of filling occurs at $p_b > 0.65$. This suggests that with further decrease in magnetic field, one might expect higher values of p_b (lower number of bonds removed) where the transition will occur. The exact form of this variation and its filling dependence would be interesting to investigate.

5.4 Summary and Future Directions

To summarize, we have studied the effect of bond percolation on the Hofstadter bands which are formed when a perpendicular magnetic field is applied to a square lattice. We have looked at the evolution of the Hofstadter butterfly as the density of removed bonds is increased. We find that at low values of p_b (= probability of having a bond), unlike the Anderson disorder, the eigenspectrum does not disperse with the magnetic field. This we attribute to the open clusters of sites which do not enclose any magnetic field. With slight increase in p_b we find few dispersing states which are due to disconnected rings. The dispersion of these are compared with finite size ring structures. At large values of p_b, we find that the disorder broadens the density of states. We also analyze the IPR of the wave functions as a function of p_b, and have looked at the effect of this disorder on band gaps and states close to $E = 0$. To understand some of the features of our results we discussed properties of finite size rings and clusters.

Next we investigated the effect of disorder on the Chern bands, and found that they undergo direct transition to the normal insulator state with increasing bond percolation disorder. This happens at a bond occupation probability p_b higher than the classical percolation threshold. We also find that the bands at the band bottom are more stable to disorder than the band center. The calculations were performed using a recently developed method of calculating Chern numbers using coupling matrix approach [51]. These results seem to be in accordance with the insights found from the diagonal Anderson disorder problem [27].

We now mention some of the future directions. In our study, we have looked at two aspects of the physics of bond percolation on square lattices when kept in presence of uniform magnetic field. One, the effect on the energy dispersion, which leads to the effective "killing" of the Hofstadter butterfly. And two, effect on transverse conductivity σ_{xy}. It will be interesting to look at the magnetic oscillations in this system for a fixed density of particles. Magnetization(M) is determined by the change of the energy dispersion of the system as a function of magnetic field $M = -\frac{\partial E}{\partial \alpha}$ [58]. If the energy spectrum does not disperse with magnetic field (α), as is the case when $p_b \ll p_c$, then this quantity will be identically *zero*. However, the exact form of this change and the variation with p_b may be interesting to investigate. Further, while we study the effect of bond percolation on the σ_{xy}, it might be interesting to correlate this with the effect on σ_{xx}. It will be intriguing to understand if the two-parameter scaling theory, as has been tested for other disorder problems in quantum Hall physics [29], is also applicable to the bond percolation disorder.

References

1. Anderson PW (1958) Absence of diffusion in certain random lattices. Phys Rev 109:1492–1505
2. Lee PA, Ramakrishnan TV (1985) Disordered electronic systems. Rev Mod Phys 57:287–337

3. Kramer B, MacKinnon A (1993) Localization: theory and experiment. Rep Prog Phys 56(12):1469
4. Janssen M (1998) Statistics and scaling in disordered mesoscopic electron systems. Phys Rep 295(12):1–91
5. Evers F, Mirlin AD (2008) Anderson transitions. Rev Mod Phys 80:1355–1417
6. Abrahams E, Kravchenko SV, Sarachik MP (2001) Metallic behavior and related phenomena in two dimensions. Rev Mod Phys 73:251–266
7. Klitzing KV, Dorda G, Pepper M (1980) New method for high-accuracy determination of the fine-structure constant based on quantized hall resistance. Phys Rev Lett 45:494–497
8. Stormer HL, Tsui DC, Gossard AC (1999) The fractional quantum hall effect. Rev Mod Phys 71:S298–S305
9. Hofstadter DR (1976) Energy levels and wave functions of bloch electrons in rational and irrational magnetic fields. Phys Rev B 14:2239–2249
10. Aidelsburger M, Atala M, Lohse M, Barreiro JT, Paredes B, Bloch I (2013) Realization of the hofstadter hamiltonian with ultracold atoms in optical lattices. Phys Rev Lett 111:185301
11. Miyake H, Siviloglou GA, Kennedy CJ, Burton WC, Ketterle W (2013) Realizing the harper hamiltonian with laser-assisted tunneling in optical lattices. Phys Rev Lett 111:185302
12. Hunt B, Sanchez-Yamagishi JD, Young AF, Yankowitz M, LeRoy BJ, Watanabe K, Taniguchi T, Moon P, Koshino M, Jarillo-Herrero P et al (2013) Massive dirac fermions and Hofstadter butterfly in a van der Waals heterostructure. Science 340(6139):1427–1430
13. Yu GL, Gorbachev RV, Tu JS, Kretinin AV, Cao Y, Jalil R, Withers F, Ponomarenko LA, Piot BA, Potemski M et al (2014) Hierarchy of Hofstadter states and replica quantum hall ferromagnetism in graphene superlattices. Nat Phys 10:525–529
14. Thouless DJ, Kohmoto M, Nightingale MP, den Nijs M (1982) Quantized hall conductance in a two-dimensional periodic potential. Phys Rev Lett 49:405–408
15. Osadchy D, Avron JE (2001) Hofstadter butterfly as quantum phase diagram. J Math Phys 42(12):5665–5671
16. Chalker JT, Coddington PD (1988) Percolation, quantum tunnelling and the integer hall effect. J Phys C: Solid State Phys 21(14):2665
17. Cain P, Römer RA, Schreiber M, Raikh ME (2001) Integer quantum hall transition in the presence of a long-range-correlated quenched disorder. Phys Rev B 64:235326
18. Galstyan AG, Raikh ME (1997) Localization and conductance fluctuations in the integer quantum hall effect: real-space renormalization-group approach. Phys Rev B 56:1422–1429
19. Kramer B, Ohtsuki T, Kettemann S (2005) Random network models and quantum phase transitions in two dimensions. Phys Rep 417(56):211–342
20. Huckestein B (1995) Scaling theory of the integer quantum hall effect. Rev Mod Phys 67:357–396
21. Ortuño M, Somoza AM, Mkhitaryan VV, Raikh ME (2011) Phase diagram of the weak-magnetic-field quantum hall transition quantified from classical percolation. Phys Rev B 84:165314
22. Dolgopolov VT (2014) Integer quantum hall effect and related phenomena. Phys-Uspekhi 57(2):105
23. Abrahams E, Anderson PW, Licciardello DC, Ramakrishnan TV (1979) Scaling theory of localization: absence of quantum diffusion in two dimensions. Phys Rev Lett 42:673–676
24. Khmelnitskii D (1984) Quantum hall effect and additional oscillations of conductivity in weak magnetic fields. Phys Lett A 106(4):182–183
25. Laughlin RB (1984) Levitation of extended-state bands in a strong magnetic field. Phys Rev Lett 52:2304–2304
26. Yang K, Bhatt RN (1996) Floating of extended states and localization transition in a weak magnetic field. Phys Rev Lett 76:1316–1319
27. Sheng DN, Weng ZY (1997) Disappearance of integer quantum hall effect. Phys Rev Lett 78:318–321
28. Pruisken AMM (1985) Dilute instanton gas as the precursor to the integral quantum hall effect. Phys Rev B 32:2636–2639

29. Sheng DN, Weng ZY (1998) New universality of the metal-insulator transition in an integer quantum hall effect system. Phys Rev Lett 80:580–583
30. Sheng DN, Weng ZY (2000) Phase diagram of the integer quantum hall effect. Phys Rev B 62:15363–15366
31. Kirkpatrick S (1973) Percolation and conduction. Rev Mod Phys 45:574–588
32. Mookerjee A, Saha-Dasgupta T, Dasgupta I (2009) Quantum transmittance through random media. In: Quantum and semi-classical percolation and breakdown in disordered solids, vol 762. Springer, Berlin, p 83
33. Koslowski T, von Niessen W (1990) Mobility edges for the quantum percolation problem in two and three dimensions. Phys Rev B 42:10342–10347
34. Islam MF, Nakanishi H (2008) Localization-delocalization transition in a two-dimensional quantum percolation model. Phys Rev E 77:061109
35. Gong L, Tong P (2009) Localization-delocalization transitions in a two-dimensional quantum percolation model: von Neumann entropy studies. Phys Rev B 80:174205
36. Dillon SB, Nakanishi H (2014) Localization phase diagram of two-dimensional quantum percolation. Eur Phys J B 87(12):1–9
37. Stauffer D, Aharony A (1991) Introduction to percolation theory. Taylor and Francis, London
38. Isichenko MB (1992) Percolation, statistical topography, and transport in random media. Rev Mod Phys 64:961–1043
39. Soukoulis CM, Grest GS (1991) Localization in two-dimensional quantum percolation. Phys Rev B 44:4685–4688
40. Odagaki T, Lax M, Puri A (1983) Hopping conduction in the d-dimensional lattice bond-percolation problem. Phys Rev B 28:2755–2765
41. Raghavan R, Mattis DC (1981) Eigenfunction localization in dilute lattices of various dimensionalities. Phys Rev B 23:4791–4793
42. Shapir Y, Aharony A, Harris AB (1982) Localization and quantum percolation. Phys Rev Lett 49:486–489
43. Taylor JPG, MacKinnon A (1989) A study of the two-dimensional bond quantum percolation model. J Phys: Condens Matter 1(49):9963
44. Schmidtke D, Khodja A, Gemmer J (2014) Transport in tight-binding bond percolation models. Phys Rev E 90:032127
45. Sanyal S, Damle K, Motrunich OI (2016) Vacancy-induced low-energy states in undoped graphene. Phys Rev Lett 117:116806
46. Häfner V, Schindler J, Weik N, Mayer T, Balakrishnan S, Narayanan R, Bera S, Evers F (2014) Density of states in graphene with vacancies: Midgap power law and frozen multifractality. Phys Rev Lett 113:186802
47. Ostrovsky PM, Protopopov IV, König EJ, Gornyi IV, Mirlin AD, Skvortsov MA (2014) Density of states in a two-dimensional chiral metal with vacancies. Phys Rev Lett 113:186803
48. Zhu L, Wang X (2016) Singularity of density of states induced by random bond disorder in graphene. Phys Lett A 380:2233–2236
49. Liu W-S, Lei X (2003) Integer quantum hall transitions in the presence of off-diagonal disorder. J Phys: Condens Matter 15(17):2693
50. Meir Y, Aharony A, Harris AB (1986) Quantum percolation in magnetic fields. Phys Rev Lett 56:976–979
51. Yi-Fu Z, Yun-You Y, Yan J, Li S, Rui S, Dong-Ning S, Ding-Yu X (2013) Coupling-matrix approach to the Chern number calculation in disordered systems. Chin Phys B 22(11):117312
52. Analytis JG, Blundell SJ, Ardavan A (2004) Landau levels, molecular orbitals, and the Hofstadter butterfly in finite systems. Am J Phys 72(5):613–618
53. Weik N, Schindler J, Bera S, Solomon GC, Evers F (2016). Graphene with vacancies: supernumerary zero modes. ArXiv e-prints, arXiv:1603.00212
54. Markoš P (2006) Numerical analysis of the anderson localization. Acta Phys Slovaca 56:561–685
55. Fradkin E (1991) Field theories of condensed matter systems, vol 7. Addison-Wesley, Redwood City

56. Niu Q, Thouless DJ, Wu Y-S (1985) Quantized hall conductance as a topological invariant. Phys Rev B 31:3372–3377
57. Dutta P, Maiti SK, Karmakar SN (2012) Integer quantum hall effect in a lattice model revisited: Kubo formalism. J Appl Phys 112(4):044306
58. Analytis JG, Blundell SJ, Ardavan A (2005) Magnetic oscillations, disorder and the Hofstadter butterfly in finite systems. Synth Metals 154(13):265–268. Proceedings of the international conference on science and technology of synthetic metals Part III

Chapter 6
Fractional Spins and Kondo Effect

6.1 Introduction

Our studies in *ill* condensed matter have been limited to noninteracting systems. In this chapter we will study an interesting interacting model —the paradigmatic Kondo effect. Here the system is rendered *ill* due to an external impurity. As was discussed in Chap. 1 Kondo effect has had a long history and many facets of this physics is now well understood. The question we address in this chapter is–what happens to Kondo effect, if the metal is spin-orbit coupled?

Systems with Rashba spin-orbit coupling(RSOC) have emerged as hosts of many new developments in condensed matter physics [1]. Examples include materia ls with topological bands [2, 3], electron gases at oxide interfaces [4, 5], cold atomic gases [6] etc. The study of quantum impurities in these systems, therefore, is of importance both from basic and applied perspectives. Some previous studies have tried to address the effect of spin-orbit coupling on Kondo effect. Malecki pointed out that Kondo effect is not destroyed by RSOC [7], and has been confirmed in a variational calculation [8]. Žitko and Bonča [9] reported that increase/decrease of T_K—the Kondo temperature, depends on microscopic parameters and attributed it to the change in density of states(DOS) engendered by RSOC. Zarea et al. reported an enhancement of T_K using an effective projected *s-d* like model [10] (see also [11–13]). In a broader context, effect of nonuniform (even diverging) DOS on impurity physics have been investigated in metals [14, 15], semiconductors [16], and superconductors [17]. A high T_K in a 2D setting was also reported [18]. Although the above discussion may suggest that quantum impurity problems with RSOC and other systems with structured DOS have been comprehensively addressed, in this chapter we demonstrate a surprising result not found in the works cited hitherto.

We will analyze a quantum impurity (with a repulsive correlation energy for double occupancy) that hybridizes with a Fermi gas with RSOC (strength λ). When RSOC and correlation energy are large enough, unlike the conventional unit local moment, we find here that the impurity develops a *fractional local moment* (fraction is 2/3). This moment couples *antiferromagnetically* with the Fermi gas, and forms a

© Springer Nature Switzerland AG 2019
A. Agarwala, *Excursions in Ill-Condensed Quantum Matter*,
Springer Theses, https://doi.org/10.1007/978-3-030-21511-8_6

Kondo like ground state where the gas screens the fractional moment. Remarkably, the resulting T_K (where the screening stops to be operative upon increase of temperature) is large—a significant fraction of the Fermi energy—and can be tuned with increasing RSOC ($T_K \sim \lambda^{4/3}$). We establish these results using a variety of methods from mean-field theory, variational approach, and quantum Monte Carlo numerics. That many body effects produce a novel fractional moment state in such a seemingly rudimentary model is itself of fundamental interest. We not only elucidate the physics of the fractional local moment in the context of RSOC systems, we further identify the essential ingredients (power of infrared divergence of DOS) in any system necessary to realize this physics (including the size of the fractional moment). We also discuss possible experimental systems that can realize these phenomena.

6.2 The System

We consider a gas ("conduction bath") of two component (spin-$\frac{1}{2}$, $\sigma = \uparrow, \downarrow$) fermions in 3D with density $n_0 \equiv \frac{k_F^3}{3\pi^2}$, and an associated Fermi energy $E_F = \frac{k_F^2}{2}$ (we set $\hbar$ and fermion mass to unity). With a RSOC of strength λ, the spin of a fermion is locked to its momentum $\boldsymbol{k}$ resulting in "helicity" $\alpha = \pm 1$ states. In terms of fermion operators $c_{k\alpha}^\dagger$, the conduction bath kinetic energy is

$$H_c = \sum_{k\alpha}(\varepsilon_\alpha(\boldsymbol{k}) - \mu)c_{k\alpha}^\dagger c_{k\alpha}, \qquad (6.1)$$

where[1] $\varepsilon_\alpha(\boldsymbol{k}) = \frac{k^2}{2} - \alpha\lambda|\boldsymbol{k}|$ and μ is the chemical potential. The spin is polarized along $\boldsymbol{k}$ for $\alpha = 1$ and opposite to $\boldsymbol{k}$ for $\alpha = -1$. Thus, $c_{k\sigma}^\dagger = \sum_\alpha f_\sigma^\alpha(\boldsymbol{k})c_{k\alpha}^\dagger$ where coefficients $f_\sigma^\alpha(\boldsymbol{k})$ are determined by $\boldsymbol{k}$. We introduce a quantum impurity d, with Hamiltonian

$$H_d = \sum_\sigma(\tilde{\varepsilon}_d - \mu)n_{d\sigma} + Un_{d\uparrow}n_{d\downarrow}, \qquad (6.2)$$

at the origin of the box of volume Ω containing the RSOC fermionic bath. Here, $n_{d\sigma} = d_\sigma^\dagger d_\sigma$, $\tilde{\varepsilon}_d$ is the "bare" impurity energy (see below), and $U(\sim$correlation energy) is the local repulsion between two $\uparrow$-$\downarrow$ fermions at the impurity. A crucial aspect is the local hybridization of the bath fermions with the impurity state given by

$$H_h = \frac{V}{\sqrt{\Omega}}\sum_{k\sigma}(c_{k\sigma}^\dagger d_\sigma + d_\sigma^\dagger c_{k\sigma}). \qquad (6.3)$$

The Hamiltonian $H = H_c + H_d + H_h$ describes the Anderson impurity problem [19] in a RSOC fermionic bath. We study the ground state and finite temperature properties of this system using various techniques. Our results (3D setting) are also applicable to other spatial dimensions (particularly 2D), and also to more general anisotropic RSOC of cold atomic systems [20].

---

[1]For convenience, a constant $\lambda^2/2$ is added to ε_α as we will see later.

6.3 Features of Free Fermi Gas with RSOC

Before we discuss the results of the complete model, it will be useful to analyze the system without the impurity. That is—what happens to the free Fermi gas with increasing spin-orbit coupling? The kinetic energy of a Fermi gas in 3D in presence of a generic Rashba spin-orbit coupling is described by the Hamiltonian

$$H_c = \int dr \Psi^\dagger(r) \left(\frac{P^2}{2} 1 - P_\lambda \cdot \tau \right) \Psi(r) \tag{6.4}$$

where the Ψ is a two component spinor

$$\Psi(r) = \begin{pmatrix} c_\uparrow(r) \\ c_\downarrow(r) \end{pmatrix}, \tag{6.5}$$

$P = -i\nabla$ is the momentum operator, 1 is SU(2) identity and τ is the vector of Pauli spin matrices τ_i, $P_\lambda = \lambda_x P_x \hat{x} + \lambda_y P_y \hat{y} + \lambda_z P_z \hat{z}$. The quantities λ_i describe a general anisotropic RSOC (which can, for example, be obtained in cold atomic systems via synthetic non-Abelian gauge fields [20]). In this work we consider a fully symmetric RSOC, $\lambda_x = \lambda_y = \lambda_z = \lambda$. Exploiting the translational symmetry, we write Hamiltonian in momentum(k) space in terms of $c_{k\sigma}$ operators which are Fourier transforms of $c_\sigma(r)$ ($\sigma = \uparrow, \downarrow$). RSOC mixes the spin $\uparrow - \downarrow$ states for a given k giving rise to helicity states with quantum numbers $\alpha = \pm 1$. The helicity bands ($\alpha = \pm 1$) have the dispersion (constant energy shift of $\lambda^2/2$ is added for convenience),

$$\varepsilon_\alpha(k) = \varepsilon_\alpha(k) = \frac{1}{2}(k - \alpha\lambda)^2, \tag{6.6}$$

where $k = |k|$. The helicity state creation operators $c_{k\alpha}^\dagger$ are related to $c_{k\sigma}^\dagger$ operators via $c_{k\sigma}^\dagger = \sum_\alpha f_\sigma^\alpha(k) c_{k\alpha}^\dagger$. Here $f_\sigma^\alpha(k) = \langle k\alpha | k\sigma \rangle$, where $|k\alpha\rangle$ are the eigenkets of RSOC Fermi gas. Explicitly $\langle k\sigma = \uparrow | k\alpha = +1 \rangle = \cos(\frac{\theta}{2})$, $\langle k\sigma = \downarrow | k\alpha = +1 \rangle = \sin(\frac{\theta}{2})e^{i\phi}$, $\langle k\sigma = \uparrow | k\alpha = -1 \rangle = -\sin(\frac{\theta}{2})$ and $\langle k\sigma = \downarrow | k\alpha = -1 \rangle = \cos(\frac{\theta}{2})e^{i\phi}$, where θ and ϕ are the polar and azimuthal angles of the vector k in spherical polar coordinates in momentum space.

RSOC Fermi gas itself (no impurity) undergoes interesting changes with changing λ. For the given density n_0, increasing λ causes the topology of the Fermi surface [20] to change at $\lambda_T = \frac{k_F}{\sqrt[3]{4}}$. For $\lambda > \lambda_T$, the Fermi sea is a spherical annulus solely of $+$ helicity fermions. For $\lambda \ll \lambda_T$, μ varies as $\frac{\mu(\lambda)}{E_F} = 1 - \frac{1}{\sqrt[3]{2}} \left(\frac{\lambda}{\lambda_T} \right)^2$, and as $\frac{\mu(\lambda)}{E_F} = \frac{2^{8/3}}{9} \left(\frac{\lambda_T}{\lambda} \right)^4$ for $\lambda \gg \lambda_T$. This transition is shown in Fig. 6.1 [21].

The bath density of states, i.e. of the RSOC fermions is given by

$$\rho(\omega) = \frac{1}{\pi^2} \left(\frac{\lambda^2}{\sqrt{2\omega}} + \sqrt{2\omega} \right) \tag{6.7}$$

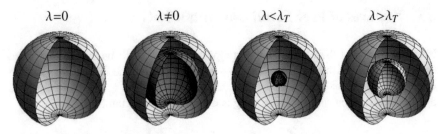

Fig. 6.1 Fermi surface topology transition. At $\lambda = 0$ the Fermi surface are two overlapping spheres both for up and down spins. Increasing λ converts spins into helicity states and the Fermi surfaces are two concentric spheres for the two distinct helicities. At $\lambda > \lambda_T$ only one kind of helicity states remain [21]

and given a density of particles ($n_0 = k_F^3/3\pi^2$), the chemical potential μ depends on λ via the relation [20]

$$\sqrt{\frac{\mu}{E_F}}\left(\frac{3\lambda^2}{k_F^2} + \frac{\mu}{E_F}\right) = 1. \tag{6.8}$$

6.4 Impurity Spectral Function

Now let us introduce an impurity in absence of the interaction term ($U = 0$). The noninteracting impurity Green's function is given by

$$\mathcal{G}_{d\sigma}(\omega) = \frac{1}{\left(\omega - \tilde{\varepsilon}_d - \sum_{k,\alpha} \frac{V^2}{2\Omega} \frac{1}{(\omega - \varepsilon_\alpha(k))}\right)}. \tag{6.9}$$

The third term in the denominator of the above expression has an ultraviolet divergence. We describe the procedure of regularization in order to handle such divergences. $\tilde{\varepsilon}_d$ is treated as a bare parameter and replaced by the corresponding physical parameter ε_d using

$$\varepsilon_d = \tilde{\varepsilon}_d - \frac{V^2}{\Omega} \sum_{|k| \leq \Lambda} \frac{1}{|k^2/2|} = \tilde{\varepsilon}_d - \frac{V^2 \Lambda}{\pi^2}. \tag{6.10}$$

All the calculations that we will present requires this important technical input. The above procedure provides a route to make all interesting observables to be independent of ultraviolet cutoff Λ.

The regularized Green's function is

$$\mathcal{G}_{d\sigma}(\omega) = \cfrac{1}{(\omega - \varepsilon_d - \left(-\cfrac{V^2\lambda^2}{2\sqrt{2\pi}\sqrt{-\omega}} + \cfrac{V^2\sqrt{-\omega}}{\sqrt{2\pi}}\right))}. \tag{6.11}$$

The corresponding impurity spectral function is

$$A_d(\omega) = 2\pi Z\delta(\omega - \varepsilon_b) + \cfrac{2\left(\cfrac{\lambda^2 V^2}{2\sqrt{2\pi}\sqrt{\omega}} + \cfrac{V^2\sqrt{\omega}}{\pi\sqrt{2}}\right)}{(\omega - \varepsilon_d)^2 + \left(\cfrac{\lambda^2 V^2}{2\sqrt{2\pi}\sqrt{\omega}} + \cfrac{V^2\sqrt{\omega}}{\pi\sqrt{2}}\right)^2} \tag{6.12}$$

where, $\frac{1}{2\pi}\int_{-\infty}^{\infty} A_d(\omega)d\omega = 1$. One notices that a bound state appears at ε_b which is the pole of the Green's function. Z is the weight of the d state in the b bound state. This, as we will see later, will be crucial in the phenomena we will discuss in the forthcoming sections. Z is evaluated by the following procedure. For any impurity Green's function of the form $\mathcal{G}_{d\sigma}(\omega) = \frac{1}{f(\omega)}$, if ε_b solves for the pole(i.e. $f(\omega = \varepsilon_b) = 0$), then $Z = \frac{1}{|f'(\omega)|}|_{\omega=\varepsilon_b}$.

6.5 Mean Field Theory

We now the consider the complete Hamiltonian and uncover the physics of this system under various approximations. We first use the Hartree-Fock (HF) approximation. In this calculation a ground state with a broken rotational symmetry is assumed, such that $M = \langle n_{d\uparrow} - n_{d\downarrow}\rangle$ is nonzero. M is self-consistently determined by minimizing the ground state energy [19]. In this method the interaction term is treated as

$$U n_{d\uparrow} n_{d\downarrow} \rightarrow U(\langle n_{d\uparrow}\rangle n_{d\downarrow} + n_{d\uparrow}\langle n_{d\downarrow}\rangle - \langle n_{d\uparrow}\rangle\langle n_{d\downarrow}\rangle). \tag{6.13}$$

The occupancy of d state for both spin labels can now be self consistently found by solving

$$\langle n_{d\sigma}\rangle = \int_{-\infty}^{\frac{\mu(\lambda)}{E_F}} \frac{-1}{\pi}\Im\left[\mathcal{G}_{d\sigma}\left(\frac{\omega^+}{E_F}, \frac{\varepsilon_d + U\langle n_{d\bar\sigma}\rangle}{E_F}, \frac{V}{E_F^{1/4}}, \frac{\lambda}{k_F}\right)\right] d\left(\frac{\omega}{E_F}\right). \tag{6.14}$$

This then allows us to find impurity moment $M = \langle n_{d\uparrow} - n_{d\downarrow}\rangle$ as a function of U and λ. Figure 6.2a shows "magnetization" M of the impurity in the U-λ plane, showing three distinct regimes. For any λ, M vanishes when $U < U_c$ ($U_c(\lambda)$ is shown by the dashed line in Fig. 6.2a. For $U > U_c$, $M \approx 1$ when $\lambda/k_F \lesssim 1$ consistent with known results [19]. Most interestingly, for $\lambda/k_F \gtrsim 1$ and $U > U_c$ we find that $M \approx 2/3$!

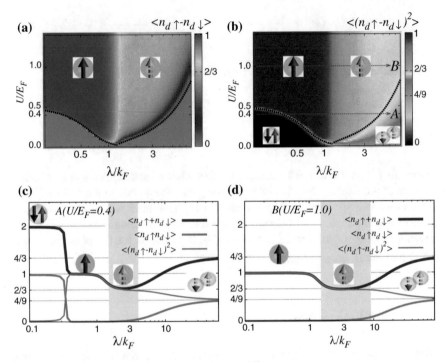

Fig. 6.2 Ground state in U-λ plane: Results for $V/E_F^{1/4} = 0.1, \varepsilon_d = \mu(\lambda)/2$. **a** Impurity moment $M = \langle n_{d\uparrow} - n_{d\downarrow}\rangle$ in the Hartree-Fock (HF) ground state. **b** Size of impurity moment $S_z^2 = \langle (n_{d\uparrow} - n_{d\downarrow})^2\rangle$ in the variational ground state. **c** and **d** Results along slices $A(U/E_F = 0.4)$ and $B(U/E_F = 1.0)$ shown in **b**. Both HF and variational ground states show a *fractional* local moment (shown schematically by the broken vector) of $2/3$ for $\lambda/k_F \gtrsim 1$, and U/E_F larger than a λ-dependent critical value shown by dashed line in **a** and **b**

6.6   Variational Calculation

To obviate any artifacts due to the artificially broken symmetry of the mean field calculation, we now construct a variational ground state (see [22]) with a "rigid" Fermi sea of bath fermions and two added particles whose spin-states are unbiased. To make this formalism numerically tenable let us first look at the resolution of identity in the basis,

$$1 = \frac{\Omega}{8\pi^3}\left(\sum_{l,m,\sigma}\int_0^\infty dk k^2 |k, l, m, \sigma\rangle\langle k, l, m, \sigma|\right) \quad (6.15)$$

where $k = |\boldsymbol{k}|$, and l, m and $\sigma = \pm 1/2$ are the azimuthal, magnetic and the spin quantum numbers respectively. $|k, l, m, \sigma\rangle$ are therefore the free particle spherical wave states. Now λ couples l, m and σ states to form *helicity* states $(l, m, \sigma) \rightarrow$

(j, m_j, α). Resolution of identity in the helicity basis is,

$$1 = \frac{\Omega}{8\pi^3} \left(\sum_{j, m_j, \alpha} \int_0^\infty dk k^2 |k, j, m_j, \alpha\rangle\langle k, j, m_j, \alpha| \right) \tag{6.16}$$

where for any k,

$$|l = 0, m = 0, \uparrow\rangle = \frac{1}{\sqrt{2}} |j = \frac{1}{2}, m_j = \frac{1}{2}, \alpha = -1\rangle - \frac{1}{\sqrt{2}} |j = \frac{1}{2}, m_j = \frac{1}{2}, \alpha = 1\rangle \tag{6.17}$$

$$|l = 0, m = 0, \downarrow\rangle = \frac{1}{\sqrt{2}} |j = \frac{1}{2}, m_j = -\frac{1}{2}, \alpha = -1\rangle - \frac{1}{\sqrt{2}} |j = \frac{1}{2}, m_j = -\frac{1}{2}, \alpha = 1\rangle. \tag{6.18}$$

The Hamiltonian H (see Eqs. (6.1), (6.2) and (6.3)) can therefore be written as,

$$\begin{aligned} H &= \sum_{j, m_j, \alpha} \frac{\Omega}{8\pi^3} \int_k k^2 dk \varepsilon_\alpha(k) |k, j, m_j, \alpha\rangle\langle k, j, m_j, \alpha| + U|d, \sigma\rangle\langle d, \sigma||d, \bar\sigma\rangle\langle d, \bar\sigma| + \sum_\sigma \tilde\varepsilon_d |d, \sigma\rangle\langle d, \sigma| \\ &+ \sum_{\sigma, \alpha} V \frac{\sqrt{\Omega}\sqrt{4\pi}}{8\pi^3} \left(\int_k k^2 dk \frac{1}{\sqrt{2}} \left(\bar\alpha |k, j = 1/2, m_j = \sigma, \alpha\rangle\langle d, \sigma| + h.c. \right) \right). \end{aligned} \tag{6.19}$$

Since $k \in (0, \infty)$, we transform $k = \tan(\frac{\pi x}{2})$ such that $dk = jac(x)dx$ where, $jac(x) = \sec^2(\frac{\pi x}{2})\frac{\pi}{2}$. The x-interval $(0, 1)$ is now further divided into discrete Gauss-Legendre points, $\int_0^1 dx \to \sum_i wt(x_i)$, such that resolution of identity (see Eq. (6.16)) can be rewritten as

$$1 = \sum_{i, j, m_j, \alpha} g(x_i) |k(x_i), j, m_j, \alpha\rangle\langle k(x_i), j, m_j, \alpha| \tag{6.20}$$

where $g(x_i) = \left(\frac{\Omega}{8\pi^3} \right) wt(x_i) jac(x_i) k(x_i)^2$. Defining $|\tilde k_i\rangle = |\tilde k(x_i)\rangle \equiv \sqrt{g(x_i)} |k(x_i)\rangle$ the complete discretized Hamiltonian is

$$\begin{aligned} H &= \sum_{i, j, m_j, \alpha} \varepsilon_\alpha(k_i) |\tilde k_i, j, m_j, \alpha\rangle\langle \tilde k_i, j, m_j, \alpha| + U n_{d\uparrow} n_{d\downarrow} + \sum_\sigma \tilde\varepsilon_d |d, \sigma\rangle\langle d, \sigma| \\ &+ V \sum_i \frac{\sqrt{2\pi g(x_i)}}{\sqrt{\Omega}} \left(-|\tilde k_i, j = \frac{1}{2}, m_j = \frac{1}{2}, \alpha = 1\rangle + |\tilde k_i, j = \frac{1}{2}, m_j = \frac{1}{2}, \alpha = -1\rangle \right) \langle d \uparrow | \\ &+ V \sum_i \frac{\sqrt{2\pi g(x_i)}}{\sqrt{\Omega}} \left(-|\tilde k_i, j = \frac{1}{2}, m_j = -\frac{1}{2}, \alpha = 1\rangle + |\tilde k_i, j = \frac{1}{2}, m_j = -\frac{1}{2}, \alpha = -1\rangle \right) \langle d \downarrow | \end{aligned} \tag{6.21}$$

The system is numerically diagonalized in the noninteracting sector ($U = 0$), where the regularization of $\tilde\varepsilon_d$ is included. A rigid Fermi sea is implemented by discarding states which have $\varepsilon_\alpha(k) < \mu(\lambda)$. The U term of the Hamiltonian is further diagonalized in the two-particle sector using the product of one particle states. Various observables can then be calculated by taking expectation on the ground state wave function. Typically $\approx 10^4$ states in the two-particle sector may be necessary to find accurate solutions.

We find that the ground state for all λ and U is rotationally invariant with a zero total (spin + orbital) angular momentum ($J = 0$, singlet). The size of the impurity local moment, characterized by $S_z^2 \equiv \langle (n_{d\uparrow} - n_{d\downarrow})^2 \rangle$, depends on λ and U. As seen from Fig. 6.2b, there are four distinct ground states. (i) For $\lambda \lesssim k_F$ and $U < U_c$ (U_c depends on λ, and is shown by a dashed line in Fig. 6.2b, S_z^2 vanishes and the impurity is doubly occupied. (ii) For $\lambda \lesssim k_F$ and $U > U_c$, $S_z^2 \simeq 1$ corresponding to the Kondo state where the impurity has a well formed local moment that locks into a singlet with the bath fermions. Interestingly, in this regime of λ, U_c falls with increasing λ, i.e. small λ aids the formation of the Kondo state(see also, [10]). The other two states occur for $\lambda \gtrsim k_F$, where U_c increases with increasing λ. (iii) For $U > U_c$, we find a strongly correlated state (vanishing double occupancy) with a *fractional* local moment of $S_z^2 = 2/3$! (iv) For $U < U_c$, an intriguing new state is seen with impurity occupancy of 4/3, moment 4/9, and double occupancy $\langle n_{d\uparrow} n_{d\downarrow} \rangle = 4/9$. The crossovers between these states are clearly demonstrated in Fig. 6.2c which shows various quantities evolving with λ for $U = 0.4 E_F$, and in Fig. 6.2d for $U = E_F$. Indeed, the HF results discussed before are consistent with those of the variational calculations (VC). We note that the first excited state of the VC is a triplet state ($J = 1$). The energy of this excited state compared to the singlet ground state gives an estimate of the Kondo scale T_K which is discussed in detail below.

6.7 Quantum Monte Carlo

Several natural questions arise including how the fractional local moment reveals itself at finite temperatures. We address this using the quantum Monte Carlo (QMC) method of Hirsch and Fye [23] which, in addition, also provides an unbiased corroboration of the results of the previous sections.

Hirsch-Fye quantum Monte Carlo numerics are performed following [23] where the susceptibility is obtained by,

$$\chi = \int_0^\beta d\tau \langle [d_\uparrow^\dagger(\tau)d_\downarrow(\tau) + d_\downarrow^\dagger(\tau)d_\uparrow(\tau)] \times [d_\uparrow^\dagger(0)d_\downarrow(0) + d_\downarrow^\dagger(0)d_\uparrow(0)] \rangle. \quad (6.22)$$

The starting Green's function can be obtained from the noninteracting impurity spectral function (see Eq. (6.12)). Throughout the calculations, the chemical potential is kept fixed at its zero-temperature value ($\mu(T, \lambda) = \mu(\lambda)$). Our formulation can be readily used to obtain quantities of interest to experiments using realistic (temperature/system dependent) values of parameters.

Figure 6.3a–c shows the temperature dependent results (including the impurity magnetic susceptibility χ) obtained from QMC for a λ and U that possesses a

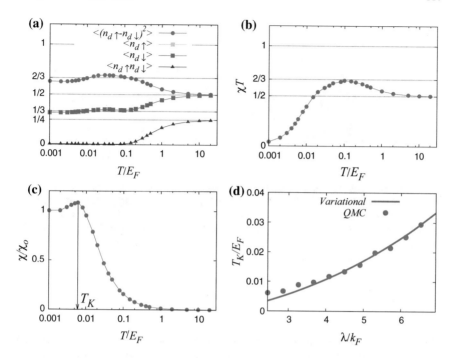

Fig. 6.3 Finite T physics: QMC results for $U/E_F = 0.5$, $\lambda/k_F = \frac{5}{\sqrt{6}}$ and $V/E_F^{1/4} = 0.1$ **a** Impurity observables, **b** and **c** impurity magnetic susceptibility χ as a function of temperature T. χ_0 in **c** is the low temperature susceptibility. Kondo temperature T_K is estimated from QMC results by the location of the peak in χ as shown in **c**. **d** Dependence of T_K on $\lambda(U/E_F = 1.0)$. Results **a**–**c** are obtained using $L = 512$ imaginary time slices, while $L = 128$ is used to obtain the T_K for various values of λ in **d**. Sampling error bars are smaller than the symbol sizes

fractional local moment in the ground state. Three temperature regimes are clearly seen. At high temperature $T \gg U$, we have the free orbital regime [24, 25] where $T\chi(T) \approx \frac{1}{2}$ (Fig. 6.3b), followed by a regime where $T\chi(T) \approx \frac{2}{3}$ at lower temperatures. At even lower temperatures (temperature scale T_K) there is a crossover to the Kondo state. The interesting aspects of these results is that the impurity local moment S_z^2 attains a value of $2/3$ in the same temperature regime where $T\chi(T) \sim \frac{2}{3}$ and remains so at low temperatures, even below the Kondo temperature T_K. This clearly indicates formation of a fractional local moment of $2/3$ at the impurity, and screening of the same by the bath fermions at lower temperatures confirming our ground state results. QMC also allows us to extract T_K as shown in Fig. 6.3c, and its dependence on λ is shown in Fig. 6.3d. The remarkable aspect is the large Kondo temperature scale that is a significant fraction of E_F, which interestingly increases with increasing λ in the fractional local moment regime. Reassuringly, the energy scale obtained from VC also agrees with the QMC result (up to a factor of $\frac{1}{2}$, $T_K^{QMC} \approx \frac{1}{2}T_K^{VC}$) as shown in Fig. 6.3d.

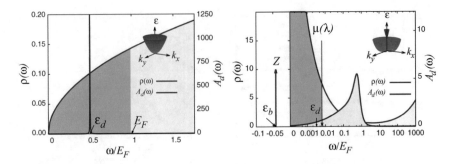

Fig. 6.4 Formation of a bound state: (Left) In absence of spin-orbit coupling the metal is described a density of states $\rho(\omega) \propto \sqrt{\omega}$. The Fermi sea is filled upto E_F. The uncoupled impurity is described by a spectral function $A_d(\omega)$ which is a δ function as $\omega = \varepsilon_d$. (Right) In presence of spin-orbit coupling, the metal density of states gets a infrared divergence of the form $\sim \frac{1}{\sqrt{\omega}}$. The chemical potential shifts to $\mu(\lambda)$. The impurity, when it now hybridizes with such a metal, forms a bound state at energy ε_b. The bound state has only a weight Z of the original impurity d

6.8   Discussion

What is the physics behind these results? In the absence of RSOC ($\lambda = 0$), the sole one-particle effect of hybridization on the impurity is to broaden its spectral function $A_d(\omega)$ from a Dirac delta at ε_d to a Lorentzian of width $\Delta \sim V^2 \rho(\mu)$ where $\rho(\omega)$ ($\sim \sqrt{\omega}$ for $\lambda = 0$) is the DOS of the bath. Matters take a different turn when $\lambda \neq 0$ due to the infrared divergence of the DOS of the bath ($\rho(\omega) \sim \frac{\lambda^2}{\sqrt{\omega}}$ at near $\omega = 0$, see Fig. 6.4). A bound state appears for *any* V for $\lambda \neq 0$, i.e. the states $\{c_{k\alpha}^\dagger, d_\sigma^\dagger\}$ reorganize themselves into a set of scattering states created by $a_{km}^\dagger$ and a one particle *bound state* $b_m^\dagger$ (quantum numbers: $k = |\mathbf{k}|$, and $m = \pm\frac{1}{2}$ is the z-projection of the total angular momentum $J = 1/2$.). In particular,

$$b_{\frac{1}{2}}^\dagger = \sqrt{Z} d_\uparrow^\dagger + \sum_{k\alpha} B_{k\alpha} c_{k\alpha}^\dagger, \qquad (6.23)$$

where Z is the weight of the d-impurity state in the bound state b, and Bs are coefficients of the bath states.

Now, Z depends on λ in a most interesting way. For a given ε_d and V, Z is vanishingly small for λ smaller than a critical value (see Fig. 6.5). For larger λ, Z attains a constant value (of $\frac{2}{3}$ for the 3D RSOC) *independent* of λ. The energy of the bound state ε_b also has interesting characteristics as shown in Fig. 6.5. For small λ, the binding energy is small and ε_d dependent, while for large λ, $\frac{\varepsilon_b}{E_F} \approx -\left(\frac{V^2\lambda^2}{k_F^3\pi}\right)^{2/3}$ and becomes independent of ε_d.

In fact important inputs can be obtained by investigating the pole of the Eq. (6.11). The value of ω which solves the pole provides the value of the bound state. When λ is small compared to other parameters the equation is solved by

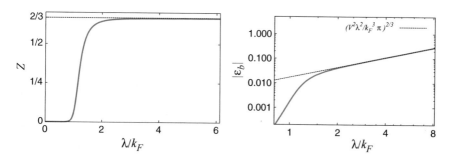

Fig. 6.5 The bound state has a fraction of the impurity: (Left) Weight(Z) of the impurity d-state in the bound state, and (Right) Energy of the bound state ε_b, as function of λ. With increasing λ while the bound state energy continues to drop in a power law fashion, the weight Z saturates to a finite fixed value = $2/3$

$$\varepsilon_b \sim -\left(\frac{V^2\lambda^2}{2\sqrt{2\pi}\,\varepsilon_d}\right)^2,\tag{6.24}$$

while when λ is large,

$$\varepsilon_b \sim -\left(\frac{V^2\lambda^2}{2\sqrt{2\pi}}\right)^{\frac{2}{3}}.\tag{6.25}$$

In this later regime Z saturates to a value of $2/3$. This is the same number which appeared as a fractional moment in all our previous investigations.

The one particle physics just discussed provides crucial clues to the physics even when U is nonzero. Clearly, the natural basis for analysis is provided by the b-bound state and the a-scattering states. For small λ, the bound state has very little d-character and the physics is quite similar to the system without RSOC. The fall in U_c seen in Fig. 6.2a, b owes to the falling chemical potential of the gas for our choice of $\varepsilon_d = \mu(\lambda)/2$. At larger λ, the bound state b is deep. Since the b state has only a fraction $\sqrt{Z}$ of d state, even a large U on the d state does *not* entirely forbid double occupancy of the b state. Physically, the part of the b state with d character will "feel" a correlation energy $Z^2 U$, while the other part remains uncorrelated. At large U, the "d-part" of b will thus be singly occupied forming a fractional local moment. This argument provides an expression for the critical U_c required to form a fractional local moment, as $U_c \sim \frac{1}{Z}|\varepsilon_b|$ and indeed matches (upto a multiplicative factor of ≈ 2) the result at large λ shown in Fig. 6.2a, b. In fact, these observations also explain the regime of $U < U_c$ at large λ. Here the b state is doubly occupied, and this corresponds to a d occupancy of $2Z$, and $\langle n_{d\uparrow}n_{d\downarrow}\rangle = Z^2$ and $S_z^2 = 2Z(1-Z)$ all in agreement with results of Fig. 6.2. The asymptotic behaviors of the bound state values from the one particle physics as shown in Eqs. (6.24) and (6.25) also provide an estimate of critical $\lambda = \lambda_c$, beyond with the fractional moment will be expected to appear. The critical value is given by

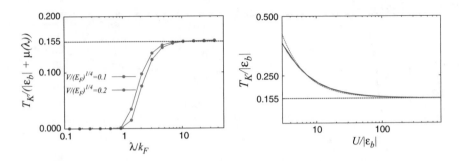

Fig. 6.6 The high Kondo temperature: (Left) The ratio of the Kondo temperature to the $\varepsilon_b + \mu(\lambda)$ is small when $\lambda \ll k_F$. However as λ increases the T_K reaches a finite fraction of ε_b which does not depend on the value of V. (Right) "Universal" Kondo T_K scale as a function of U estimated from the variational calculation where $\frac{T_K}{|\varepsilon_b|} \approx 0.155 + \frac{|\varepsilon_b|}{U}$

$$\frac{\lambda_c}{k_F} = \frac{1}{\sqrt{3}} \left(\frac{3\pi}{2V^2/\sqrt{E_F}} \right)^{1/8}. \tag{6.26}$$

Turning again to $U > U_c$, the origin of the high T_K of the Kondo state formed by the fractional local moment can be understood from the variational calculation. As noted, the first excited state in VC is a triplet state made of a *singly occupied b* state and a scattering state at the chemical potential. This state is clearly a scale ε_b above the ground state. Thus in the large U limit we expect the Kondo scale T_K to be proportional to ε_b as indeed found by explicit calculation (see Fig. 6.6). This provides a route to obtain large Kondo temperatures as $T_K \sim \lambda^{4/3}$. Also note that the physics of the fractional local moment formation in this system is very different from that noted in Ref. [26] which occurs in a *s-d* system that has a *ferromagnetic* coupling to the bath.

Finally, why is Z numerically equal to $\frac{2}{3}$? What controls this? Can it be tuned? We show that Z is *entirely determined by the exponent that characterizes the infrared divergence of the density of states*, independent of details such as spatial dimensions. For a system with

$$\rho(\omega) = \frac{1}{\pi^2} \left(\sqrt{2\omega} + \frac{\lambda}{\sqrt{2}} \frac{\omega^r}{\lambda^{2r}} \right) \tag{6.27}$$

The infrared divergence is characterized by the exponent r $(-1 < r < 0)$. For a given density of particles n_o, one can obtain the dependence of μ on both r and λ (similar to Eq. (6.8)). The impurity Green's function and Z can be again obtained. It is found that for large λ, $Z \to \frac{1}{1-r}$. This can be obtained analytically, by discarding the $\sim \sqrt{\omega}$ term in $\rho(\omega)$ and considering $\rho(\omega) = \omega^r$ $(-1 < r < 0)$. The impurity Green's function in this case is given by,

$$G_d(\omega) = \frac{1}{\omega - \varepsilon_d - V^2 \pi (-\frac{1}{\omega})^{-r} \csc(\pi r)} \tag{6.28}$$

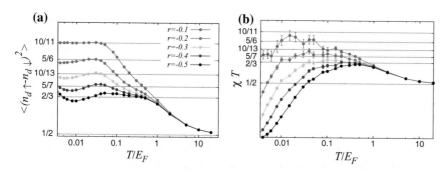

Fig. 6.7 Generic fractional local moments: QMC results for an impurity hybridizing with a conduction bath with $\rho(\omega) = \frac{1}{\pi^2}(\sqrt{2\omega} + \frac{\lambda}{\sqrt{2}}\frac{\omega^r}{\lambda^{2r}})$, $(V/E_F^{1/4} = 0.1, \lambda/k_F = \frac{1}{\sqrt{2}}(\frac{10}{\sqrt{3}})^{\frac{2}{1-2r}}, U/E_F = 2, \varepsilon_d = \mu_r(\lambda)/2, L = 128)$ for different values of r. Impurity observables (a) and susceptibility χ (b) as a function of temperature. The values of $\varepsilon_b/E_F \approx -0.1$ for all cases

with $Z = \frac{1}{1-r}$ for all values of $V(\neq 0)$ and $\varepsilon_d = 0$.

We have performed QMC calculations with the impurity hybridizing to a bath with the given DOS in Eq. (6.27), and indeed find the anticipated fractional local moments (see Fig. 6.7a). We further see (Fig. 6.7b) that there are two distinct intermediate temperature regimes, $T_K \lesssim T \lesssim |\varepsilon_b|$ which is the "fractional local moment regime" with $T\chi \approx Z(r)$, and the asymmetric local moment regime between $|\varepsilon_b| \lesssim T \lesssim U$ where $T\chi \approx \frac{2}{3}$. Interestingly, for the 3D RSOC the susceptibility alone cannot discern these two.

6.9 Perspective

This chapter unequivocally establishes that in presence of strong spin-orbit coupling one could in fact realize a "fractional spin". While the understanding of this phenomena can be uncovered by examining one-particle physics, the nontrivial aspect is the formation of the fractional Kondo effect! All diagnostics which can be employed to examine a moment exhibit a fractional value. That a simple model such as this can exhibit such a remarkable emergent phenomena is by itself interesting; however, it will be most exciting to realize a fractional moment experimentally. Possible experimental pursuits include those in cold atoms, where by combining approaches described in Ref. [27] for the 3D RSOC, and [28] for the impurity (see also [29–31]) one could possibly realize this physics. Signatures of the fractional local moment formation can be probed using radio-frequency(rf) spectroscopy [32] on the impurity. A finite concentration of well-separated quantum impurities would show a well-separated peak in the rf spectrum of ↑ spin, proportional to the concentration and the weight Z. The bath states make up a fraction of $(1 - Z) \sim \frac{1}{3}$ of the bound state, with an equal quantum mechanical admixture of ↑ and ↓ states of the bath. If the rf spectrum of the bath atoms is probed, this will again result in a well separated peak (at the bound state energy) $\propto (1 - Z)$. A combination of these two measurements can provide evidence for the physics discussed here. Material realizations of such physics will be also be noteworthy.

A final comment: in this chapter we have investigated a three dimensional model, however, analogous physics applies to 2D system with RSOC which has a similar infrared divergence of $\rho(\omega)(\sim \frac{1}{\sqrt{\omega}})$. Such 2D systems with strong spin-orbit coupling have been realized in interfaces [4, 33] and surfaces [34] where RSOC scale is comparable to the Fermi energy. It might be interesting to explore ways to realize the physics of fractional local moment in these systems as well.

References

1. Manchon A, Koo HC, Nitta J, Frolov SM, Duine RA (2015) New perspectives for Rashba spin-orbit coupling. Nat Mater 14:871–882 (Review)
2. Hasan MZ, Kane CL (2010) Colloquium: topological insulators. Rev Mod Phys 82:3045–3067
3. Qi X-L, Zhang S-C (2011) Topological insulators and superconductors. Rev Mod Phys 83:1057–1110
4. Caviglia AD, Gabay M, Gariglio S, Reyren N, Cancellieri C, Triscone J-M (2010) Tunable Rashba spin-orbit interaction at oxide interfaces. Phys Rev Lett 104:126803
5. Coey J, Ariando N, Pickett W (2013) Magnetism at the edge: new phenomena at oxide interfaces. MRS Bull 38:1040–1047
6. Goldman N, Juzeliūnas G, Öhberg P, Spielman IB (2014) Light-induced gauge fields for ultracold atoms. Rep Prog Phys 77(12):126401
7. Malecki J (2007) The two dimensional Kondo model with Rashba spin-orbit coupling. J Stat Phys 129(4):741–757
8. Feng X-Y, Zhang F-C (2011) Kondo spin screening cloud in two-dimensional electron gas with spinorbit couplings. J Phys: Condens Matter 23(10):105602
9. Žitko R, Bonča J (2011) Kondo effect in the presence of Rashba spin-orbit interaction. Phys Rev B 84:193411
10. Zarea M, Ulloa SE, Sandler N (2012) Enhancement of the Kondo effect through Rashba spin-orbit interactions. Phys Rev Lett 108:046601
11. Isaev L, Agterberg DF, Vekhter I (2012) Kondo effect in the presence of spin-orbit coupling. Phys Rev B 85:081107
12. Yanagisawa T (2012) Kondo effect in the presence of spinorbit coupling. J Phys Soc Jpn 81(9):094713
13. Chen L, Sun J, Tang H-K, Lin H-Q (2015). The Kondo temperature of a two-dimensional electron gas with Rashba spin-orbit coupling. ArXiv e-prints, arXiv:1503:00449
14. Gonzalez-Buxton C, Ingersent K (1998) Renormalization-group study of Anderson and Kondo impurities in gapless fermi systems. Phys Rev B 57:14254–14293
15. Mitchell AK, Vojta M, Bulla R, Fritz L (2013) Quantum phase transitions and thermo-dynamics of the power-law Kondo model. Phys Rev B 88:195119
16. Galpin MR, Logan DE (2008) Anderson impurity model in a semiconductor. Phys Rev B 77:195108
17. Balatsky AV, Vekhter I, Zhu J-X (2006) Impurity-induced states in conventional and unconventional superconductors. Rev Mod Phys 78:373–433
18. Wong A, Ulloa SE, Sandler N, Ingersent K (2016) Influence of Rashba spin-orbit coupling on the Kondo effect. Phys Rev B 93:075148
19. Anderson PW (1961) Localized magnetic states in metals. Phys Rev 124:41–53

20. Vyasanakere JP, Zhang S, Shenoy VB (2011) BCS-BEC crossover induced by a synthetic non-abelian gauge field. Phys Rev B 84:014512
21. Vyasanekere JP (2013) Ultracold fermions in synthetic non-abelian gauge fields. Thesis, Indian Institute of Science
22. Yosida K (1966) Bound state due to the s-d exchange interaction. Phys Rev 147:223–227
23. Hirsch JE, Fye RM (1986) Monte carlo method for magnetic impurities in metals. Phys Rev Lett 56:2521–2524
24. Krishna-murthy HR, Wilkins JW, Wilson KG (1980) Renormalization-group approach to the Anderson model of dilute magnetic alloys. i. static properties for the symmetric case. Phys Rev B 21:1003–1043
25. Krishna-murthy HR, Wilkins JW, Wilson KG (1980) Renormalization-group approach to the Anderson model of dilute magnetic alloys. ii. static properties for the asymmetric case. Phys Rev B 21:1044–1083
26. Vojta M, Bulla R (2002) A fractional-spin phase in the power-law Kondo model. Eur Phys J B-Condens Matter Complex Syst 28(3):283–287
27. Anderson BM, Juzeliūnas G, Galitski VM, Spielman IB (2012) Synthetic 3D spin-orbit coupling. Phys Rev Lett 108:235301
28. Bauer J, Salomon C, Demler E (2013) Realizing a Kondo-correlated state with ultracold atoms. Phys Rev Lett 111:215304
29. Nishida Y (2013) Su (3) orbital Kondo effect with ultracold atoms. Phys Rev Lett 111:135301
30. Falco GM, Duine RA, Stoof HTC (2004) Molecular Kondo resonance in atomic fermi gases. Phys Rev Lett 92:140402
31. Kuzmenko I, Kuzmenko T, Avishai Y, Kikoin K (2015) Model for overscreened Kondo effect in ultracold fermi gas. Phys Rev B 91:165131
32. Ketterle W, Zwierlein MW (2008) Making, probing and understanding ultracold fermi gases. Nuovo Cimento Rivista Serie 31:247–422
33. Joshua A, Ruhman J, Pecker S, Altman E, Ilani S (2013) Gate-tunable polarized phase of two-dimensional electrons at the LaAlO$_3$/SrTiO$_3$ interface. Proc Natl Acad Sci 110(24):9633–9638
34. Santander-Syro AF, Fortuna F, Bareille C, Rödel TC, Landolt G, Plumb NC, Dil JH, Radović M (2014) Giant spin splitting of the two-dimensional electron gas at the surface of SrTiO$_3$. Nat Mater 13:1085–1090

Chapter 7
Structure of Many-Body Hamiltonians in Different Symmetry Classes

Among the various *ill* systems we have analyzed upto now, most were noninteracting, apart from the study in previous chapter where we looked at the effect of an correlated impurity in a spin-orbit coupled system. This chapter provides the framework to analyze generic interacting Hamiltonians with arbitrary interactions. It looks at the essential interplay of non-ordinary symmetries as introduced in Chap. 1 and the resulting constraints on the structure of many-body fermionic Hamiltonians. As we will see, the analysis in this chapter will provide us the recipe to construct many-body Hamiltonians in any of the ten symmetry classes.

The many-body dynamics of a fermionic system, comprising of L orbitals as introduced in Chap. 1, is determined by the Hamiltonian which contains up to N-body interactions where $0 \leqslant N \leqslant L$, and is formally written as

$$\mathscr{H} = \sum_{K=0}^{N} (\Psi^\dagger)^K \mathbf{H}^{(K)} (\Psi)^K \tag{7.1}$$

where

$$
\begin{aligned}
(\Psi^\dagger)^K \mathbf{H}^{(K)} (\Psi)^K &\equiv H^{(K)}_{i_1 i_2 \dots i_K; j_1 j_2 \dots j_K} \left(\psi_{i_1} \psi_{i_2} \dots \psi_{i_K} \right)^\dagger \psi_{j_1} \psi_{j_2} \dots \psi_{j_K} \\
&= H^{(K)}_{i_1 i_2 \dots i_K; j_1 j_2 \dots j_K} \psi^\dagger_{i_K} \psi^\dagger_{i_{K-1}} \dots \psi^\dagger_{i_1} \psi_{j_1} \psi_{j_2} \dots \psi_{j_K} \tag{7.2}
\end{aligned}
$$

with repeated indices is and js summed from $1, \dots, L$. Note that here we depart slightly from the usual convention for the many-body Hamiltonian where $\dagger$ operation is done on the complete string of ψs; this notation will eventually prove to be useful in the later manipulations. Note also that $\left(\psi_{i_1} \psi_{i_2} \dots \psi_{i_K} \right)^\dagger$ should be distinguished from the definition of the $1 \times L$ dimensional array $\Psi^\dagger$ in Eq. (2.7). $\mathbf{H}^{(K)}$ is the matrix which describes the K-body interactions, and its components $H^{(K)}_{i_1 i_2 \dots i_K; j_1 j_2 \dots j_K}$ have two properties. First, $H^{(K)}_{i_1 i_2 \dots i_K; j_1 j_2 \dots j_K}$ is fully antisymmetric under permutations of the i indices among themselves, and also under the permutations of the j indices among themselves. Expressed in an equation

© Springer Nature Switzerland AG 2019
A. Agarwala, *Excursions in Ill-Condensed Quantum Matter*,
Springer Theses, https://doi.org/10.1007/978-3-030-21511-8_7

$$H^{(K)}_{i_{X(1)}i_{X(2)}\ldots i_{X(K)};j_{Y(1)}j_{Y(2)}\ldots j_{Y(K)}} = \text{sgn}(X)\,\text{sgn}(Y)\,H^{(K)}_{i_1 i_2\ldots i_K;j_1 j_2\ldots j_K} \qquad (7.3)$$

where X and Y are arbitrary permutations of K objects, and sgn denotes the sign of the permutation. Second, the Hermitian character of the Hamiltonian is reflected in the relation

$$H^{(K)}_{j_1 j_2\ldots j_K;i_1 i_2\ldots i_K} = \left(H^{(K)}_{i_1 i_2\ldots i_K;j_1 j_2\ldots j_K}\right)^* . \qquad (7.4)$$

Each $\mathbf{H}^{(K)}$ is an element of an $\binom{L}{K}$-dimensional vector space $\mathcal{H}^{(K)}$ over the *real* numbers $\mathbb{R}$. With no further restrictions other than Eqs. (7.3) and (7.4), in fact, this vector space is endowed with a structure of a *Lie algebra*. This is achieved by constructing an isomorphic vector space $\mathrm{i}\mathcal{H}^{(K)}$ (multiplying every element of $\mathcal{H}^{(K)}$ by $\mathrm{i} = \sqrt{-1}$). It is evident that for any two matrices $\mathrm{i}\mathbf{H}_a^{(K)}$ and $\mathrm{i}\mathbf{H}_b^{(K)}$ of $\mathrm{i}\mathcal{H}^{(K)}$, the commutator $[\mathrm{i}\mathbf{H}_a^{(K)}, \mathrm{i}\mathbf{H}_b^{(K)}]$ is also a matrix in $\mathrm{i}\mathcal{H}^{(K)}$, and in fact,

$$\mathrm{i}\mathcal{H}^{(K)} \sim \mathbf{u}\left(\binom{L}{K}\right), \qquad (7.5)$$

i.e., $\mathrm{i}\mathcal{H}^{(K)}$ is isomorphic to the well known Lie algebra $\mathbf{u}\left(\binom{L}{K}\right)$ which generates the Lie group $U\left(\binom{L}{K}\right)$ of $\binom{L}{K} \times \binom{L}{K}$ dimensional unitary matrices. The Hamiltonian Eq. (7.1) can be described by a $N+1$ tuple

$$\mathbf{H} = (\mathbf{H}^{(0)}, \mathbf{H}^{(1)}, \ldots, \mathbf{H}^{(N)}) \in \mathcal{H} \qquad (7.6)$$

where $\mathcal{H}$ is the real vector space

$$\mathcal{H} = \bigtimes_{K=0}^{N} \mathcal{H}^{(K)} \qquad (7.7)$$

The problem of classification of a fermionic system of L-orbitals and upto N-body interactions can now be stated precisely. How many "distinct" spaces $\mathcal{H}$ are possible? The symmetries of the system will determine the distinct structures of these spaces, placing them in different classes.

7.1 Symmetry Conditions

A symmetry operation is a symmetry if it leaves the Hamiltonian of the system invariant. For usual symmetries, this is effected by the condition

$$\mathscr{U}_{\text{USL}} \mathscr{H} \mathscr{U}_{\text{USL}}^{-1} = \mathscr{H} \qquad (7.8)$$

while for the transposing symmetry operation the condition changes to

$$: \mathscr{U}_{\mathrm{TRN}} \mathscr{H} \mathscr{U}_{\mathrm{TRN}}^{-1} := \mathscr{H} \qquad (7.9)$$

where : : indicates that the expression $\mathscr{U}_{\mathrm{TRN}} \mathscr{H} \mathscr{U}_{\mathrm{TRN}}^{-1}$ has to be normal ordered (all creation operators to the left of annihilation operators) using the anticommutation relations Eq. (2.5). Both of these types of symmetries induces a mapping of $\mathbf{H}$ in Eq. (7.6) to $\mathring{\mathbf{H}}$ via

$$\mathbf{H} = (\mathbf{H}^{(0)}, \mathbf{H}^{(1)}, \ldots, \mathbf{H}^{(N)}) \mapsto \mathring{\mathbf{H}} = (\mathring{\mathbf{H}}^{(0)}, \mathring{\mathbf{H}}^{(1)}, \ldots, \mathring{\mathbf{H}}^{(N)}). \qquad (7.10)$$

To obtain $\mathring{\mathbf{H}}^{(K)}$ for any K, we introduce an intermediate quantity $\check{\mathbf{H}}^{(K)}$ which is determined by the type of symmetry operation (see Fig. 2.2):

$$\check{H}^{(K)}_{i_1,\ldots,i_K;j_1,\ldots,j_K} = \begin{cases} U_{i_1 i'_1} \cdots U_{i_K i'_K} H^{(K)}_{i'_1,\ldots,i'_K;j'_1,\ldots,j'_K} U^{\dagger}_{j'_1 j_1} \cdots U^{\dagger}_{j'_K j_K}, & \mathrm{UL} \\ U_{i_1 i'_1} \cdots U_{i_K i'_K} \left(H^{(K)}_{i'_1,\ldots,i'_K;j'_1,\ldots,j'_K} \right)^* U^{\dagger}_{j'_1 j_1} \cdots U^{\dagger}_{j'_K j_K}, & \mathrm{UA} \\ U^*_{i_1 i'_1} \cdots U^*_{i_K i'_K} H^{(K)}_{i'_1,\ldots,i'_K;j'_1,\ldots,j'_K} U^{T}_{j'_1 j_1} \cdots U^{T}_{j'_K j_K}, & \mathrm{TL} \\ U^*_{i_1 i'_1} \cdots U^*_{i_K i'_K} \left(H^{(K)}_{i'_1,\ldots,i'_K;j'_1,\ldots,j'_K} \right)^* U^{T}_{j'_1 j_1} \cdots U^{T}_{j'_K j_K}, & \mathrm{TA} \end{cases} \qquad (7.11)$$

Here all the primed indices are summed from 1 to L, and $\mathbf{U}$s are the unitary matrices that encode the symmetry operations. Note that antilinear symmetry operations lead to a complex conjugation of the matrix elements. The transformation of the Hamiltonian Eq. (7.10) can now be specified completely. For usual symmetries, both linear and antilinear, we have

$$\mathring{\mathbf{H}}^{(K)} = \check{\mathbf{H}}^{(K)}. \qquad (7.12)$$

For transposing operation, the result is a bit more involved on account of the normal ordering operation of Eq. (7.9). We find

$$\mathring{\mathbf{H}}^{(K)} = \sum_{R=K}^{N} A_{R,K} \left[\mathrm{tr}_{R-K} \check{\mathbf{H}}^{(R)} \right]^{T}, \qquad (7.13)$$

The trace operation tr is defined as

$$\mathrm{tr}_P \check{\mathbf{H}}^{(K)} \equiv \check{H}^{(K)}_{i_1 i_2 \ldots i_{(K-P)} k_1 k_2 \ldots k_P; j_1 j_2 \ldots j_{(K-P)} k_1 k_2 \ldots k_P} \qquad (7.14)$$

accomplishing the tracing out (repeated k indices are summed) of P indices in $\check{\mathbf{H}}^{(K)}$, resulting in a $\binom{L}{K-P} \times \binom{L}{K-P}$ matrix. The constants $A_{K,R}$ in Eq. (7.13) are computed to be

$$A_{R,K} = \begin{cases} (-1)^K \dfrac{1}{(R-K)!} \left(\dfrac{R!}{K!} \right)^2, & 0 \leqslant K \leqslant R \\ 0, & K > R \end{cases} \qquad (7.15)$$

Note that the RHS of Eq. (7.13) involves a matrix transpose, and this is a characteristic of the transposing symmetry operations justifying our terminology. Thus, Eqs. (7.10), (7.11), (7.12) and (7.13) completely determine the mapping of $\mathbf{H}$ to $\mathring{\mathbf{H}}$. A symmetry operation (of any type) describes a symmetry if and only if

$$\mathring{\mathbf{H}} = \mathbf{H} \tag{7.16}$$

which is a concise expression of Eqs. (7.8) and (7.9).

7.2 Framework for Systems with Interactions

In this section we establish ideas that allow for the determination of the structure of Hamiltonians with upto N-body interactions as in Eq. (7.1). For usual symmetries, the conditions that determine the spaces $\mathfrak{i}\mathcal{H}^{(K)}$ are straightforward. From, Eq. (7.12) we obtain

$$\mathbf{H}^{(K)} = \check{\mathbf{H}}^{(K)} . \tag{7.17}$$

These are a set of homogeneous equations in the matrix elements $H^{(K)}_{i_1 \dots i_K; j_1 \dots j_K}$ that make $\mathfrak{i}\mathcal{H}^{(K)}$ of the appropriate class a vector subspace of $\mathbf{u}\left(\binom{L}{K}\right)$ following the discussion near Eq. (7.5). Equation (7.17) provides conditions to completely determine the structure of the admissible Hamiltonians.

Transposing symmetries have a more involved story. To make progress, we rewrite Eq. (7.16) using Eq. (7.13) to obtain

$$\begin{aligned}
\mathbf{H}^{(K)} &= \sum_{R=K}^{N} A_{R,K} \left[\mathrm{tr}_{R-K} \check{\mathbf{H}}^{(R)} \right]^{T} \\
&= (-1)^{K} \left[\check{\mathbf{H}}^{(K)} \right]^{T} + (-1)^{K} \sum_{R>K}^{N} \frac{1}{(R-K)!} \left(\frac{R!}{K!} \right)^{2} \left[\mathrm{tr}_{R-K} \check{\mathbf{H}}^{(R)} \right]^{T}
\end{aligned} \tag{7.18}$$

We now define $\mathbf{H}^{(K)}_{+}$ and $\mathbf{H}^{(K)}_{-}$ as

$$\begin{aligned}
\mathbf{H}^{(K)}_{+} &\equiv \mathbf{H}^{(K)} + (-1)^{K} \left[\check{\mathbf{H}}^{(K)} \right]^{T} \\
\mathbf{H}^{(K)}_{-} &\equiv \mathbf{H}^{(K)} - (-1)^{K} \left[\check{\mathbf{H}}^{(K)} \right]^{T}
\end{aligned} \tag{7.19}$$

with $\mathbf{H}^{(K)} = \frac{1}{2}\left(\mathbf{H}^{(K)}_{+} + \mathbf{H}^{(K)}_{-}\right)$ and $(-1)^{K}\left[\check{\mathbf{H}}^{(K)}\right]^{T} = \frac{1}{2}\left(\mathbf{H}^{(K)}_{+} - \mathbf{H}^{(K)}_{-}\right)$. With these definitions Eq. (7.13) becomes

$$\mathbf{H}_{-}^{(K)} = \frac{(-1)^K}{2} \sum_{R>K}^{N} \frac{(-1)^R}{(R-K)!} \left(\frac{R!}{K!}\right)^2 \left[\mathrm{tr}_{R-K}\mathbf{H}_{+}^{(R)} - \mathrm{tr}_{R-K}\mathbf{H}_{-}^{(R)}\right]. \tag{7.20}$$

Two points emerge from these considerations: (i) the transposing symmetry condition Eq. (7.18) puts no constraint on $\mathbf{H}_{+}^{(K)}$ for any K, and (ii) for every K, $\mathbf{H}_{-}^{(K)}$ is solely determined by R-body interactions terms where $R > K$. This implies that the set of independent parameters of $\mathbf{H}$ is entirely determined by $\mathbf{H}_{+}^{(K)}$ for K from 0 to N. Further note that every $\mathbf{H}_{+}^{(K)}$ belongs to a vector subspace of $\mathfrak{u}\left(\binom{L}{K}\right)$ defined by $\mathbf{H}^{(K)} = \mathbf{0}$, which we call $\mathbf{i}\mathcal{H}_{+}^{(K)}$. Since, $\mathbf{i}\mathcal{H}^{(K)}$ is made of objects of the type $\frac{1}{2}\left(\mathbf{H}_{+}^{(K)} + \mathbf{H}_{-}^{(K)}\right)$ where $\mathbf{H}_{-}^{(K)}$ is a "constant" determined by Eq. (7.20), $\mathbf{i}\mathcal{H}^{(K)} = \frac{\mathbf{i}}{2}\left(\mathbf{H}_{+}^{(K)} + \mathbf{H}_{-}^{(K)}\right)$, is no longer a vector subspace of $\mathfrak{u}\left(\binom{L}{K}\right)$. In fact, $\mathbf{i}\mathcal{H}^{(K)}$ is an affine subspace of $\mathfrak{u}\left(\binom{L}{K}\right)$ whose dimension (as a manifold) is same as the subspace $\mathbf{i}\mathcal{H}_{+}^{(K)}$.

These observations show that $\mathcal{H}$ in any class can be completely determined by starting from the N-body (highest multi-body interaction) term in $\mathbf{H}$ which satisfies $\mathbf{H}_{-}^{(N)} = \mathbf{0}$, and recursively using Eq. (7.20) to determine $\mathbf{H}_{-}^{(K)}$ for $K < N$. Again, for each K the equation $\mathbf{H}_{-}^{(K)} = \mathbf{0}$ gives the subspace that defines $\mathbf{i}\mathcal{H}_{+}^{(K)}$ which then makes up the affine space $\mathbf{i}\mathcal{H}^{(K)}$. The problem of finding the structure of $\mathcal{H}$ is then reduced solely to finding the subspace $\mathbf{i}\mathcal{H}_{+}^{(K)}$ for each K. In the subsequent sections we shall demonstrate the determination of this subspace $\mathbf{H}_{-}^{(N)} = \mathbf{0}$ for the highest multi-body (N-body) interaction in our system. Clearly, the same results will apply mutatis mutandis to all $K < N$. Finally, we note that $\mathbf{H}^{(R)}$ s satisfy the identity

$$\frac{1}{2}\sum_{R>0}^{N}(-1)^R R! \left[\mathrm{tr}_R\mathbf{H}_{+}^{(R)} - \mathrm{tr}_R\mathbf{H}_{-}^{(R)}\right] = 0. \tag{7.21}$$

In the next section we show how this is achieved for $N = 2$, and generalize this to higher N in the subsequent sections. The main results of this exercise are summarized in Tables 7.1 and 7.2.

7.3 Structure of Two-Body Hamiltonians

In this section we find the structure of two-body Hamiltonians in each class. This will serve as a precursor to our study of a generic even N-body Hamiltonian. We define a *strictly* two-body Hamiltonian as

$$\mathcal{H} = \Psi^{\dagger}\Psi^{\dagger}\mathbf{H}^{(2)}\Psi\Psi, \tag{7.22}$$

where $\Psi\Psi$ is a notational short-hand for a column vector comprised of $\binom{L}{2}$ distinct product terms of two fermionic annihilation operators, written as

Table 7.1 Structure of N-body interaction hamiltonian $\mathbf{H}^{(N)}$ (N even) in each symmetry class. The space of K-body (K even) interaction Hamiltonians $i\mathcal{H}^{(K)}$ in a class is an affine subspace $i\mathcal{H}^{(K)} = i\mathcal{H}_+^{(K)} + \mathbf{H}_-^{(K)}$ where $i\mathcal{H}_+^{(K)}$ is to be read from the last column of this table and $\mathbf{H}_-^{(K)}$ is given in Eq. (7.20)

Class	L	P	Q	$\mathbf{H}^{(N)}$	$\dim i\mathcal{H}^{(N)}$	$i\mathcal{H}_+^N$
A (0,0,0)	L	$\binom{L}{N}$	–	$\mathbf{H}^{(N)} = [\mathbf{H}^{(N)}]^\dagger$	P^2	$\mathfrak{u}(P)$
AI (+1,0,0)	L	$\binom{L}{N}$	–	$\mathbf{H}^{(N)} = [\mathbf{H}^{(N)}]^*$	$P(P+1)/2$	$\mathfrak{u}(P) \setminus \mathfrak{o}(P)$
AII (−1,0,0)	$L = 2M$	$\frac{1}{2}\left\{\binom{L}{N} + \binom{M}{N/2}\right\}$	$\frac{1}{2}\left\{\binom{L}{N} - \binom{M}{N/2}\right\}$	$\begin{pmatrix} \mathbf{h}_{PP}^{(N)} & \mathbf{h}_{PQ}^{(N)} \\ [\mathbf{h}_{PQ}^{(N)}]^\dagger & \mathbf{h}_{QQ}^{(N)} \end{pmatrix}$ $\mathbf{h}_{PP}^{(N)} = [\mathbf{h}_{PP}^{(N)}]^*$ $\mathbf{h}_{QQ}^{(N)} = [\mathbf{h}_{QQ}^{(N)}]^*$ $\mathbf{h}_{PQ}^{(N)} = -[\mathbf{h}_{PQ}^{(N)}]^*$	$\frac{P(P+1)}{2} + \frac{Q(Q+1)}{2} + PQ$	$\mathfrak{u}(P+Q)$ $\setminus \mathfrak{o}(P+Q)$
D (0,+1,0)	L	$\binom{L}{N}$	–	$\mathbf{H}^{(N)} = [\mathbf{H}^{(N)}]^*$	$P(P+1)/2$	$\mathfrak{u}(P+Q)$ $\setminus \mathfrak{o}(P)$
C (0,−1,0)	$L = 2M$	$\frac{1}{2}\left\{\binom{L}{N} + \binom{M}{N/2}\right\}$	$\frac{1}{2}\left\{\binom{L}{N} - \binom{M}{N/2}\right\}$	$\begin{pmatrix} \mathbf{h}_{PP}^{(N)} & \mathbf{h}_{PQ}^{(N)} \\ [\mathbf{h}_{PQ}^{(N)}]^\dagger & \mathbf{h}_{QQ}^{(N)} \end{pmatrix}$ $\mathbf{h}_{PP}^{(N)} = [\mathbf{h}_{PP}^{(N)}]^*$ $\mathbf{h}_{QQ}^{(N)} = [\mathbf{h}_{QQ}^{(N)}]^*$ $\mathbf{h}_{PQ}^{(N)} = -[\mathbf{h}_{PQ}^{(N)}]^*$	$\frac{P(P+1)}{2} + \frac{Q(Q+1)}{2} + PQ$	$\mathfrak{u}(P+Q)$ $\setminus \mathfrak{o}(P+Q)$
AIII (0,0,1)	$L = p+q$	$\sum_{a=1,3,\ldots}^{N-1}\binom{p}{a}\binom{q}{N-a}$	$\sum_{a=0,2,\ldots}^{N}\binom{p}{a}\binom{q}{N-a}$	$\begin{pmatrix} \mathbf{h}_{PP}^{(N)} & 0_{PQ} \\ 0_{QP} & \mathbf{h}_{QQ}^{(N)} \end{pmatrix}$ $\mathbf{h}_{PP}^{(N)} = [\mathbf{h}_{PP}^{(N)}]^\dagger$ $\mathbf{h}_{QQ}^{(N)} = [\mathbf{h}_{QQ}^{(N)}]^\dagger$	$P^2 + Q^2$	$\mathfrak{u}(P) \oplus \mathfrak{u}(Q)$
BDI (+1,+1,1)	$L = p+q$ $p = 2r\ q = 2s$	$\sum_{a=1,3,\ldots}^{N-1}\binom{p}{a}\binom{q}{N-a}$	$\sum_{a=0,2,\ldots}^{N}\binom{p}{a}\binom{q}{N-a}$	$\begin{pmatrix} \mathbf{h}_{PP}^{(N)} & 0_{PQ} \\ 0_{QP} & \mathbf{h}_{QQ}^{(N)} \end{pmatrix}$ $\mathbf{h}_{PP}^{(N)} = [\mathbf{h}_{PP}^{(N)}]^*$ $\mathbf{h}_{QQ}^{(N)} = [\mathbf{h}_{QQ}^{(N)}]^*$	$\frac{P(P+1)}{2} + \frac{Q(Q+1)}{2}$	$(\mathfrak{u}(P) \setminus \mathfrak{o}(P))$ $\oplus$ $(\mathfrak{u}(Q) \setminus \mathfrak{o}(Q))$
CII (−1,−1,1)	$L = p+q$ $p = 2r\ q = 2s$	$P = \sum_{a=1,3,\ldots}^{N-1}\binom{p}{a}\binom{q}{N-a}$ $A(B) = P/2$	$Q = \sum_{a=0,2,\ldots}^{N}\binom{p}{a}\binom{q}{N-a}$ $C(D) = \frac{Q}{2} \pm \sum_{a=0,2,\ldots}^{N}\frac{1}{2}\binom{r}{a/2}\binom{s}{(N-a)/2}$	$\begin{pmatrix} \mathbf{h}_{AA}^{(N)} & \mathbf{h}_{AB}^{(N)} \\ [\mathbf{h}_{AB}^{(N)}]^\dagger & \mathbf{h}_{BB}^{(N)} \end{pmatrix} \quad 0_{PQ}$ $0_{QP} \quad \begin{pmatrix} \mathbf{h}_{CC}^{(N)} & \mathbf{h}_{CD}^{(N)} \\ [\mathbf{h}_{CD}^{(N)}]^\dagger & \mathbf{h}_{DD}^{(N)} \end{pmatrix}$ $\mathbf{h}_{AA(CC)}^{(N)} = [\mathbf{h}_{AA(CC)}^{(N)}]^*$ $\mathbf{h}_{BB(DD)}^{(N)} = [\mathbf{h}_{BB(DD)}^{(N)}]^*$ $\mathbf{h}_{AB(CD)}^{(N)} = -[\mathbf{h}_{AB(CD)}^{(N)}]^*$	$\frac{A(A+1)}{2} + \frac{B(B+1)}{2} + AB$ $+ \frac{C(C+1)}{2} + \frac{D(D+1)}{2} + CD$	$(\mathfrak{u}(A+B)$ $\setminus \mathfrak{o}(A+B))$ $\oplus$ $(\mathfrak{u}(C+D)$ $\setminus \mathfrak{o}(C+D))$
CI (+1,−1,1)	$L = 2M$	$P = \sum_{a=1,3,\ldots}^{N-1}\binom{M}{a}\binom{M}{N-a}$ $A(B) = \begin{cases} P/2 & : N/2 \text{ even} \\ \frac{P}{2} \pm \frac{1}{2}\binom{M}{N/2} & : N/2 \text{ odd} \end{cases}$	$Q = \sum_{a=0,2,\ldots}^{N}\binom{M}{a}\binom{M}{N-a}$ $C(D) = \begin{cases} \frac{Q}{2} \pm \frac{1}{2}\binom{M}{N/2} & : N/2 \text{ even} \\ Q/2 & : N/2 \text{ odd} \end{cases}$	$\begin{pmatrix} \mathbf{h}_{AA}^{(N)} & \mathbf{h}_{AB}^{(N)} \\ [\mathbf{h}_{AB}^{(N)}]^\dagger & \mathbf{h}_{BB}^{(N)} \end{pmatrix} \quad 0_{PQ}$ $0_{QP} \quad \begin{pmatrix} \mathbf{h}_{CC}^{(N)} & \mathbf{h}_{CD}^{(N)} \\ [\mathbf{h}_{CD}^{(N)}]^\dagger & \mathbf{h}_{DD}^{(N)} \end{pmatrix}$ $\mathbf{h}_{AA(CC)}^{(N)} = [\mathbf{h}_{AA(CC)}^{(N)}]^*$ $\mathbf{h}_{BB(DD)}^{(N)} = [\mathbf{h}_{BB(DD)}^{(N)}]^*$ $\mathbf{h}_{AB(CD)}^{(N)} = [\mathbf{h}_{AB(CD)}^{(N)}]^*$	$\frac{A(A+1)}{2} + \frac{B(B+1)}{2} + AB$ $+ \frac{C(C+1)}{2} + \frac{D(D+1)}{2} + CD$	-do-
DIII (−1,+1,1)	$L = 2M$	$P = \sum_{a=1,3,\ldots}^{N-1}\binom{M}{a}\binom{M}{N-a}$ $A(B) = \begin{cases} P/2 & : N/2 \text{ even} \\ \frac{P}{2} \pm \frac{1}{2}\binom{M}{N/2} & : N/2 \text{ odd} \end{cases}$	$Q = \sum_{a=0,2,\ldots}^{N}\binom{M}{a}\binom{M}{N-a}$ $C(D) = \begin{cases} \frac{Q}{2} \pm \frac{1}{2}\binom{M}{N/2} & : N/2 \text{ even} \\ Q/2 & : N/2 \text{ odd} \end{cases}$	$\begin{pmatrix} \mathbf{h}_{AA}^{(N)} & \mathbf{h}_{AB}^{(N)} \\ [\mathbf{h}_{AB}^{(N)}]^\dagger & \mathbf{h}_{BB}^{(N)} \end{pmatrix} \quad 0_{PQ}$ $0_{QP} \quad \begin{pmatrix} \mathbf{h}_{CC}^{(N)} & \mathbf{h}_{CD}^{(N)} \\ [\mathbf{h}_{CD}^{(N)}]^\dagger & \mathbf{h}_{DD}^{(N)} \end{pmatrix}$ $\mathbf{h}_{AA(CC)}^{(N)} = [\mathbf{h}_{AA(CC)}^{(N)}]^*$ $\mathbf{h}_{BB(DD)}^{(N)} = [\mathbf{h}_{BB(DD)}^{(N)}]^*$ $\mathbf{h}_{AB(CD)}^{(N)} = -[\mathbf{h}_{AB(CD)}^{(N)}]^*$	$\frac{A(A+1)}{2} + \frac{B(B+1)}{2} + AB$ $+ \frac{C(C+1)}{2} + \frac{D(D+1)}{2} + CD$	-do-

Table 7.2 The ten symmetry classes of fermions and the structure of many-body Hamiltonians $\mathbf{H}^{(N)}$ when N is odd. The space of K-body (K odd) interaction Hamiltonians $i\mathcal{H}^{(K)}$ in a class is an affine subspace $i\mathcal{H}^{(K)} = i\mathcal{H}_+^{(K)} + \mathbf{H}_-^{(K)}$ where $i\mathcal{H}_+^{(K)}$ is to be read from the last column of this table and $\mathbf{H}_-^{(K)}$ is given in Eq. (7.20)

Class	L	P	Q	$\mathbf{H}^{(N)}$	dim $i\mathcal{H}^{(N)}$	$i\mathcal{H}_+^{(N)}$
A (0,0,0)	L	$\binom{L}{N}$	—	$\mathbf{H}^{(N)} = \left[\mathbf{H}^{(N)}\right]^\dagger$	P^2	$\mathfrak{u}(P)$
AI (+1,0,0)	L	$\binom{L}{N}$	—	$\mathbf{H}^{(N)} = \left[\mathbf{H}^{(N)}\right]^*$	$P(P+1)/2$	$\mathfrak{u}(P)\setminus\mathfrak{o}(P)$
AII (−1,0,0)	$L = 2M$	$\frac{1}{2}\binom{2M}{N}$	$\frac{1}{2}\binom{2M}{N}$	$\begin{pmatrix} \mathbf{h}_{PP}^{(N)} & \mathbf{h}_{PQ}^{(N)} \\ -\left[\mathbf{h}_{PQ}^{(N)}\right]^* & \left[\mathbf{h}_{PP}^{(N)}\right]^* \end{pmatrix}$	$P^2 + 2\times\frac{P(P-1)}{2}$	$\mathfrak{u}(2P)\setminus\mathfrak{usp}(2P)$
D (0,+1,0)	L	$\binom{L}{N}$	—	$\mathbf{H}^{(N)} = -\left[\mathbf{H}^{(N)}\right]^*$	$P(P-1)/2$	$\mathfrak{o}(P)$
C (0,−1,0)	$L = 2M$	$\frac{1}{2}\binom{2M}{N}$	$\frac{1}{2}\binom{2M}{N}$	$\begin{pmatrix} \mathbf{h}_{PP}^{(N)} & \mathbf{h}_{PQ}^{(N)} \\ \left[\mathbf{h}_{PQ}^{(N)}\right]^* & -\left[\mathbf{h}_{PP}^{(N)}\right]^* \end{pmatrix}$	$P^2 + 2\times\frac{P(P+1)}{2}$	$\mathfrak{usp}(2P)$
AIII (0,0,1)	$L = p+q$	$\sum_{a=1,3,\dots}^{N}\binom{p}{a}\binom{q}{N-a}$	$\sum_{a=0,2,\dots}^{N-1}\binom{p}{a}\binom{q}{N-a}$	$\begin{pmatrix} \mathbf{0}_{PP} & \mathbf{h}_{PQ}^{(N)} \\ \left[\mathbf{h}_{PQ}^{(N)}\right]^\dagger & \mathbf{0}_{QQ} \end{pmatrix}$	$2PQ$	$\mathfrak{u}(P+Q)\setminus(\mathfrak{u}(P)\oplus\mathfrak{u}(Q))$
BDI (+1,+1,1)	$L = p+q$	$\sum_{a=1,3,\dots}^{N}\binom{p}{a}\binom{q}{N-a}$	$\sum_{a=0,2,\dots}^{N-1}\binom{p}{a}\binom{q}{N-a}$	$\begin{pmatrix} \mathbf{0}_{PP} & \mathbf{h}_{PQ}^{(N)} \\ \left[\mathbf{h}_{PQ}^{(N)}\right]^T & \mathbf{0}_{QQ} \end{pmatrix}$	PQ	$\mathfrak{o}(P+Q)\setminus(\mathfrak{o}(P)\oplus\mathfrak{o}(Q))$
CII (−1,−1,1)	$L = p+q$ $p = 2r \quad q = 2s$	$P = \sum_{a=1,3,\dots}^{N}\binom{p}{a}\binom{q}{N-a}$ $A(B) = \frac{P}{2}$	$Q = \sum_{a=0,2,\dots}^{N-1}\binom{p}{a}\binom{q}{N-a}$ $C(D) = \frac{Q}{2}$	$\begin{pmatrix} \mathbf{0}_{PP} & \begin{matrix} \mathbf{h}_{AC}^{(N)} & \mathbf{h}_{AD}^{(N)} \\ -\left[\mathbf{h}_{AD}^{(N)}\right]^* & \left[\mathbf{h}_{AC}^{(N)}\right]^* \end{matrix} \\ \text{h.c.} & \mathbf{0}_{QQ} \end{pmatrix}$	PQ	$\mathfrak{usp}(P+Q)\setminus(\mathfrak{usp}(P)\oplus\mathfrak{usp}(Q))$
CI (+1,−1,1)	$L = 2M$	$\frac{1}{2}\binom{2M}{N}$	$\frac{1}{2}\binom{2M}{N}$	$\begin{pmatrix} \mathbf{0}_{PP} & \mathbf{h}_{PQ}^{(N)} \\ \left[\mathbf{h}_{PQ}^{(N)}\right]^* & \mathbf{0}_{QQ} \end{pmatrix}$	$P(P+1)$	$\mathfrak{usp}(2P)\setminus\mathfrak{u}(P)$
DIII (−1,+1,1)	$L = 2M$	$\frac{1}{2}\binom{2M}{N}$	$\frac{1}{2}\binom{2M}{N}$	$\begin{pmatrix} \mathbf{0}_{PP} & \mathbf{h}_{PQ}^{(N)} \\ -\left[\mathbf{h}_{PQ}^{(N)}\right]^* & \mathbf{0}_{QQ} \end{pmatrix}$	$P(P-1)$	$\mathfrak{o}(2P)\setminus\mathfrak{u}(P)$

$$\Psi\Psi \equiv \begin{pmatrix} L-1 \begin{cases} \psi_1\psi_2 \\ \vdots \\ \psi_1\psi_L \end{cases} \\ L-2 \begin{cases} \psi_2\psi_3 \\ \vdots \\ \psi_3\psi_L \end{cases} \\ \vdots \\ 1 \begin{cases} \psi_{L-1}\psi_L \end{cases} \end{pmatrix}. \tag{7.23}$$

Henceforth, we will call these product terms as states. The definition of $\Psi\Psi$ also defines $\Psi^\dagger\Psi^\dagger \equiv$

$$\left((\psi_1\psi_2)^\dagger \dots (\psi_1\psi_L)^\dagger \; (\psi_2\psi_3)^\dagger \dots (\psi_2\psi_L)^\dagger \; \dots \; (\psi_{L-1}\psi_L)^\dagger \right). \tag{7.24}$$

The preceding definitions are in the same spirit as our definition for the states in Eq. (2.7).

We can also write $\mathbf{H}^{(2)}$ as (see Eq. (7.2))

$$(\Psi^\dagger)^2 \mathbf{H}^{(2)} (\Psi)^2 = \sum_{i_1, i_2, j_1, j_2} (\psi_{i_1}\psi_{i_2})^\dagger H^{(2)}_{i_1, i_2; j_1, j_2} \psi_{j_1}\psi_{j_2} \tag{7.25}$$

Note here that although $\mathbf{H}^{(2)}$ is a four indexed object, fermionic anticommutation demands that it is a matrix of dimension $\binom{L}{2} \times \binom{L}{2}$. As done in the case of one-body Hamiltonians we now construct the $\binom{L}{2} \times \binom{L}{2}$ canonical representations of the symmetry operations($\mathbf{U}_T^{(2)}, \mathbf{U}_C^{(2)}, \mathbf{U}_S^{(2)}$) from their one-body counterparts(see Table 2.1) and apply them to determine the structure of $\mathbf{H}^{(2)}$.

Class A: This class has no symmetries, therefore any hermitian matrix of dimension $\binom{L}{2} \times \binom{L}{2}$ belongs to this class.

Class AI: For the AI class we have $\mathbf{U}_T = \mathbf{1}$. Hence for this case,

$$\mathscr{T}\Psi^\dagger\mathscr{T}^{-1} = \Psi^\dagger\mathbf{U}_T = \Psi^\dagger \qquad \mathscr{T}\Psi\mathscr{T}^{-1} = \mathbf{U}_T^\dagger\Psi = \Psi$$

and for the two-body states (see Eqs. (7.23) and (7.24)) we have

$$\mathscr{T}\Psi^\dagger\Psi^\dagger\mathscr{T}^{-1} = \Psi^\dagger\Psi^\dagger \qquad \mathscr{T}\Psi\Psi\mathscr{T}^{-1} = \Psi\Psi, \tag{7.26}$$

making $\mathbf{U}_T^{(2)} = \mathbf{1}$. Time reversal symmetry implies the following

$$\mathbf{H}^{(2)} = \left[\mathbf{H}^{(2)} \right]^* \tag{7.27}$$

which puts $\mathbf{H}^{(2)}$ in the class of real symmetric $\binom{L}{2} \times \binom{L}{2}$ matrices. The dimension of this class is $\frac{1}{2}\binom{L}{2}\left\{\binom{L}{2}+1\right\}$.

Class AII: For class AII we have canonical $\mathbf{U}_T = \mathbf{J}$ and

$$\mathcal{T}\Psi^\dagger \mathcal{T}^{-1} = \Psi^\dagger \mathbf{U}_T = \Psi^\dagger \mathbf{J}$$
$$\mathcal{T}\Psi \mathcal{T}^{-1} = \mathbf{U}_T^\dagger \Psi = -\mathbf{J}\Psi. \tag{7.28}$$

Also we had seen that for this case to be realizable $L = 2M$. We can divide the $2M$ states as M states of flavor α and M of flavor β. Explicitly these transform in the following way

$$\mathcal{T}\left(\psi_{1\alpha}^\dagger \cdots \psi_{M\alpha}^\dagger \ \psi_{1\beta}^\dagger \cdots \psi_{M\beta}^\dagger\right)\mathcal{T}^{-1}$$

$$= \left(-\psi_{1\beta}^\dagger \cdots - \psi_{M\beta}^\dagger \ \psi_{1\alpha}^\dagger \cdots \psi_{M\alpha}^\dagger\right)$$

$$\mathcal{T}\begin{pmatrix} \psi_{1\alpha} \\ \vdots \\ \psi_{M\alpha} \\ \psi_{1\beta} \\ \vdots \\ \psi_{M\beta} \end{pmatrix}\mathcal{T}^{-1} = \begin{pmatrix} -\psi_{1\beta} \\ \vdots \\ -\psi_{M\beta} \\ \psi_{1\alpha} \\ \vdots \\ \psi_{M\alpha} \end{pmatrix} \tag{7.29}$$

implying,

$$\mathcal{T}\psi_{i\alpha}^\dagger \mathcal{T}^{-1} = -\psi_{i\beta}^\dagger \qquad \mathcal{T}\psi_{i\beta}^\dagger \mathcal{T}^{-1} = \psi_{i\alpha}^\dagger$$
$$\mathcal{T}\psi_{i\alpha} \mathcal{T}^{-1} = -\psi_{i\beta} \qquad \mathcal{T}\psi_{i\beta} \mathcal{T}^{-1} = \psi_{i\alpha}. \tag{7.30}$$

Since $\Psi\Psi$ is formed by the product of these one-body operators, its elements can be divided into the following three kinds

$$\Psi\Psi = \begin{pmatrix} \binom{M}{2}\left\{\psi_{i\alpha}\psi_{j\alpha}\right. \\ M^2\left\{\psi_{i\alpha}\psi_{j\beta}\right. \\ \binom{M}{2}\left\{\psi_{i\beta}\psi_{j\beta}\right. \end{pmatrix}. \tag{7.31}$$

The number of states in each kind add upto $\binom{L}{2}$, i.e., $2\binom{M}{2} + M^2 = M(M-1) + M^2 = M(2M-1) = \frac{L(L-1)}{2} = \binom{L}{2}$. We find the effect of the $\mathcal{T}$ operators on these states to be

$$\mathscr{T}\Psi\Psi\,\mathscr{T}^{-1} = \mathscr{T}\begin{pmatrix} \psi_{i\alpha}\psi_{j\alpha} \\ \psi_{i\alpha}\psi_{j\beta} \\ \psi_{i\beta}\psi_{j\beta} \end{pmatrix}\mathscr{T}^{-1} = \begin{pmatrix} \psi_{i\beta}\psi_{j\beta} \\ -\psi_{i\beta}\psi_{j\alpha} \\ \psi_{i\alpha}\psi_{j\alpha} \end{pmatrix} = \begin{pmatrix} \psi_{i\beta}\psi_{j\beta} \\ \psi_{j\alpha}\psi_{i\beta} \\ \psi_{i\alpha}\psi_{j\alpha} \end{pmatrix}. \quad (7.32)$$

This suggests that we can re-organize the two-body basis into "symmetric" (σ) and "antisymmetric" (π) states as follows[1]

$$\begin{pmatrix} \binom{M}{2}\left\{\psi_{i\alpha}\psi_{j\alpha}\right. \\ M^2\left\{\psi_{i\alpha}\psi_{j\beta}\right. \\ \binom{M}{2}\left\{\psi_{i\beta}\psi_{j\beta}\right. \end{pmatrix} \longrightarrow \begin{pmatrix} \sigma\begin{cases} \binom{M}{2}\left\{\psi_{i\alpha}\psi_{j\alpha}+\psi_{i\beta}\psi_{j\beta}\right. \\ M\left\{\psi_{i\alpha}\psi_{i\beta}\right. \\ \binom{M}{2}\left\{\psi_{i\alpha}\psi_{j\beta}+\psi_{j\alpha}\psi_{i\beta}\ (i\neq j)\right. \end{cases} \\ \pi\begin{cases} \binom{M}{2}\left\{\psi_{i\alpha}\psi_{j\beta}-\psi_{j\alpha}\psi_{i\beta}\ (i\neq j)\right. \\ \binom{M}{2}\left\{\psi_{i\alpha}\psi_{j\alpha}-\psi_{i\beta}\psi_{j\beta}\right. \end{cases} \end{pmatrix}. \quad (7.33)$$

Conveniently, in this way of organizing our Hilbert space, we have

$$\mathscr{T}\sigma\,\mathscr{T}^{-1} = \sigma \qquad \mathscr{T}\pi\,\mathscr{T}^{-1} = -\pi. \quad (7.34)$$

Also note that $M^2 = 2\binom{M}{2} + M$.

It will be useful to mention here that the transformation properties of these states under symmetry gives us the definitions of "symmetric" and "antisymmetric" states. This terminology will be used in later sections when the N-body Hamiltonians will be discussed. Note that the word "symmetric" may not necessarily imply a "+" sign in the linear combination of its constituent states and vice-versa.

Armed with this, we look for the structure of $\mathbf{H}^{(2)}$

$$\mathscr{T}\Psi^\dagger\Psi^\dagger\mathbf{H}^{(2)}\Psi\Psi\,\mathscr{T}^{-1} = \Psi^\dagger\Psi^\dagger\mathbf{H}^{(2)}\Psi\Psi$$
$$\mathbf{U}_T^{(2)}\left[\mathbf{H}^{(2)}\right]^*\mathbf{U}_T^{(2)\dagger} = \mathbf{H}^{(2)}, \quad (7.35)$$

where $\mathbf{U}_T^{(2)}$ in this new basis is obtained from Eq. (7.34) and is given by

$$\left[\mathbf{U}_T^{(2)}\right]^\dagger = \mathbf{1}_{P,Q}, \quad (7.36)$$

where $P = 2\binom{M}{2} + M$ and $Q = 2\binom{M}{2}$. Assuming,

$$\mathbf{H}^{(2)} = \begin{pmatrix} \mathbf{h}_{PP}^{(2)} & \mathbf{h}_{PQ}^{(2)} \\ \left[\mathbf{h}_{PQ}^{(2)}\right]^\dagger & \mathbf{h}_{QQ}^{(2)} \end{pmatrix} \quad (7.37)$$

---

[1] For the clarity of the equations, we will not explicitly put the factors of $\frac{1}{\sqrt{2}}$ in front of symmetric and antisymmetric states. It should be assumed that all the states are properly normalized.

and implementing the condition in Eq. (7.35), we get the following constraints,

$$\mathbf{h}_{PP}^{(2)} = \left[\mathbf{h}_{PP}^{(2)}\right]^*, \quad \mathbf{h}_{QQ}^{(2)} = \left[\mathbf{h}_{QQ}^{(2)}\right]^*, \quad \mathbf{h}_{PQ}^{(2)} = -\left[\mathbf{h}_{PQ}^{(2)}\right]^*. \tag{7.38}$$

This leads to total $\frac{P(P+1)}{2} + \frac{Q(Q+1)}{2} + PQ$ independent parameters in $\mathbf{H}^{(2)}$, which is equal to dim $i\mathcal{H}_{All}^{(2)}$.

Class D: We now look at transposing symmetries of the TL kind. The states transform according to Eq. (2.23) and since class D has canonical $\mathbf{U_C} = \mathbf{1}$, we have

$$\mathcal{U}\Psi^\dagger\mathcal{U}^{-1} = \Psi^T \qquad \mathcal{U}\Psi\mathcal{U}^{-1} = (\Psi^\dagger)^T \tag{7.39}$$

which implies

$$\mathcal{U}\psi_i^\dagger\mathcal{U}^{-1} = \psi_i \qquad \mathcal{U}\psi_i\mathcal{U}^{-1} = \psi_i^\dagger. \tag{7.40}$$

Thus even for the two-body Hamiltonian the symmetry condition (Eq. (2.24)) is

$$\mathbf{H}^{(2)} = \left[\mathbf{H}^{(2)}\right]^*, \tag{7.41}$$

making $\mathbf{H}^{(2)}$ a real symmetric matrix with $\frac{1}{2}\binom{L}{2}\left\{\binom{L}{2}+1\right\}$ parameters.

Class C: Class C comes with canonical $\mathbf{U_C} = \mathbf{J}$ and requires $L = 2M$. The states transform as

$$\mathcal{U}\Psi^\dagger\mathcal{U}^{-1} = \Psi^T\mathbf{J} \qquad \mathcal{U}\Psi\mathcal{U}^{-1} = -\mathbf{J}(\Psi^\dagger)^T. \tag{7.42}$$

Similar to class All we label the states with α, β flavors and find there transformations to be

$$\begin{aligned}
\mathscr{C}\psi_{i\alpha}^\dagger\mathscr{C}^{-1} &= -\psi_{i\beta} & \mathscr{C}\psi_{i\beta}^\dagger\mathscr{C}^{-1} &= \psi_{i\alpha} \\
\mathscr{C}\psi_{i\alpha}\mathscr{C}^{-1} &= -\psi_{i\beta}^\dagger & \mathscr{C}\psi_{i\beta}\mathscr{C}^{-1} &= \psi_{i\alpha}^\dagger.
\end{aligned} \tag{7.43}$$

However, unlike class All, one must remember that here we are dealing with a transposing symmetry. The two-body basis in this case can be rearranged in the following manner

$$\begin{pmatrix} \binom{M}{2}\left\{\psi_{i\alpha}\psi_{j\alpha}\right. \\ M^2\left\{\psi_{i\alpha}\psi_{j\beta}\right. \\ \binom{M}{2}\left\{\psi_{i\beta}\psi_{j\beta}\right. \end{pmatrix} \longrightarrow \begin{pmatrix} \pi \begin{cases} \binom{M}{2}\left\{\psi_{i\alpha}\psi_{j\alpha} + \psi_{i\beta}\psi_{j\beta}\right. \\ M\left\{\psi_{i\alpha}\psi_{i\beta}\right. \\ \binom{M}{2}\left\{\psi_{i\alpha}\psi_{j\beta} + \psi_{j\alpha}\psi_{i\beta}\ (i\neq j)\right. \end{cases} \\ \sigma \begin{cases} \binom{M}{2}\left\{\psi_{i\alpha}\psi_{j\beta} - \psi_{j\alpha}\psi_{i\beta}\ (i\neq j)\right. \\ \binom{M}{2}\left\{\psi_{i\alpha}\psi_{j\alpha} - \psi_{i\beta}\psi_{j\beta}\right. \end{cases} \end{pmatrix}. \tag{7.44}$$

The action of $\mathscr{C}$ symmetry on these states are

$$\mathscr{C}\sigma\mathscr{C}^{-1} = \sigma^\dagger \qquad \mathscr{C}\pi\mathscr{C}^{-1} = -\pi^\dagger. \tag{7.45}$$

From this we get $2\binom{M}{2}$ number of σ states and M^2 number of π states. Imposing $\mathscr{C}$ symmetry on the Hamiltonian gives

$$: \mathscr{C}\Psi^\dagger\Psi^\dagger\mathbf{H}^{(2)}\Psi\Psi\mathscr{C}^{-1} : = \Psi^\dagger\Psi^\dagger\mathbf{H}^{(2)}\Psi\Psi$$
$$\mathbf{U}_C^{(2)\dagger}\left[\mathbf{H}^{(2)}\right]^* \mathbf{U}_C^{(2)} = \mathbf{H}^{(2)} \tag{7.46}$$

where

$$\mathbf{U}_C^{(2)} = -\mathbf{1}_{P,Q} \tag{7.47}$$

and $P = 2\binom{M}{2} + M$ and $Q = 2\binom{M}{2}$. Denoting the internal structure of $\mathbf{H}^{(2)}$ with Eq. (7.37), the constraints turn out to be same as Eq. (7.38). This again leads to $\frac{P(P+1)}{2} + \frac{Q(Q+1)}{2} + PQ$ independent parameters.

Class AIII: For transposing symmetries of the TA kind, we know

$$: (\mathscr{U}\mathscr{H}\mathscr{U}^{-1}) : = \mathscr{H}. \tag{7.48}$$

Class AIII has canonical $\mathbf{U}_S = \mathbf{1}_{p,q}$ and

$$\mathscr{U}\Psi^\dagger\mathscr{U}^{-1} = \Psi^T\mathbf{1}_{p,q} \qquad \mathscr{U}\Psi\mathscr{U}^{-1} = \mathbf{1}_{p,q}(\Psi^\dagger)^T. \tag{7.49}$$

Here, $L = p + q$. We label p-type states with α and the q-type states with β and they transform as

$$\mathscr{U}\psi_{i\alpha}^\dagger\mathscr{U}^{-1} = \psi_{i\alpha} \qquad \mathscr{U}\psi_{i\beta}^\dagger\mathscr{U}^{-1} = -\psi_{i\beta}$$
$$\mathscr{U}\psi_{i\alpha}\mathscr{U}^{-1} = \psi_{i\alpha}^\dagger \qquad \mathscr{U}\psi_{i\beta}\mathscr{U}^{-1} = -\psi_{i\beta}^\dagger. \tag{7.50}$$

The two-body states look like

$$\Psi\Psi = \begin{pmatrix} \binom{p}{2}\left\{\psi_{i\alpha}\psi_{j\alpha}\right\} \\ pq\left\{\psi_{i\alpha}\psi_{j\beta}\right\} \\ \binom{q}{2}\left\{\psi_{i\beta}\psi_{j\beta}\right\} \end{pmatrix}. \tag{7.51}$$

Also note that $\binom{p}{2} + pq + \binom{q}{2} = \binom{L}{2}$ and

$$\mathscr{S}\Psi\Psi\mathscr{S}^{-1} = \mathscr{S}\begin{pmatrix} \psi_{i\alpha}\psi_{j\alpha} \\ \psi_{i\alpha}\psi_{j\beta} \\ \psi_{i\beta}\psi_{j\beta} \end{pmatrix}\mathscr{S}^{-1} = \begin{pmatrix} -(\psi_{i\alpha}\psi_{j\alpha})^\dagger \\ (\psi_{i\alpha}\psi_{j\beta})^\dagger \\ -(\psi_{i\beta}\psi_{j\beta})^\dagger \end{pmatrix}, \tag{7.52}$$

which shows that terms with odd and even αs will transform differently under the $\mathscr{S}$ symmetry. We reorganize the states as

$$
\begin{pmatrix} \psi_{i\alpha}\psi_{j\alpha} \\ \psi_{i\alpha}\psi_{j\beta} \\ \psi_{i\beta}\psi_{j\beta} \end{pmatrix} \rightarrow \begin{pmatrix} P \begin{Bmatrix} \psi_{i\alpha}\psi_{j\beta} \end{Bmatrix} \\ Q \begin{Bmatrix} \psi_{i\alpha}\psi_{j\alpha} \\ \psi_{i\beta}\psi_{j\beta} \end{Bmatrix} \end{pmatrix}, \tag{7.53}
$$

where $P = pq$ and $Q = \binom{p}{2} + \binom{q}{2}$. Symmetry then demands

$$
: \mathscr{S}\Psi^\dagger\Psi^\dagger\mathbf{H}^{(2)}\Psi\Psi\mathscr{S}^{-1} : = \Psi^\dagger\Psi^\dagger\mathbf{H}^{(2)}\Psi\Psi
$$
$$
\mathbf{U}_S^{(2)\dagger}\mathbf{H}^{(2)}\mathbf{U}_S^{(2)} = \mathbf{H}^{(2)} \tag{7.54}
$$

where

$$
\mathbf{U}_S^{(2)} = \mathbf{1}_{P,Q}. \tag{7.55}
$$

Using the structure of $\mathbf{H}^{(2)}$ as given in Eq. (7.37) and imposing Eq. (7.54), the constraints come out as

$$
\mathbf{h}_{PQ}^{(2)} = \mathbf{0}, \quad \mathbf{h}_{PP}^{(2)} = \left[\mathbf{h}_{PP}^{(2)}\right]^\dagger, \quad \mathbf{h}_{QQ}^{(2)} = \left[\mathbf{h}_{QQ}^{(2)}\right]^\dagger. \tag{7.56}
$$

The off-diagonal blocks of $\mathbf{H}^{(2)}$ are forced to be *zero*, which is in contrast with structure of the one-body Hamiltonian in class AIII(see Table 2.2). The number of independent parameters are $P^2 + Q^2$.

Before delving into the classes BDI, CII, CI and DIII, we bring to the attention of the reader that since all of them respect the sublattice symmetry (S = 1), the Hamiltonian for all four classes will be of the form

$$
\mathbf{H}^{(2)} = \begin{pmatrix} \mathbf{h}_{PP}^{(2)} & \mathbf{0} \\ \mathbf{0} & \mathbf{h}_{QQ}^{(2)} \end{pmatrix}. \tag{7.57}
$$

The presence of T = ±1 and C = ±1 symmetries will only put constraints on the nonzero sub-blocks of $\mathbf{H}^{(2)}$.

Class BDI: This class has $\mathbf{U}_S = \mathbf{1}_{p,q}$, $\mathbf{U}_C = \mathbf{1}_{p,q}$, $\mathbf{U}_T = \mathbf{1}$. Since time reversal symmetry operation has a canonical $\mathbf{U}_T = \mathbf{1}$, we now have additional conditions on the sub-blocks (see Eq. (7.57)),

$$
\mathbf{h}_{PP}^{(2)} = \left[\mathbf{h}_{PP}^{(2)}\right]^*, \quad \mathbf{h}_{QQ}^{(2)} = \left[\mathbf{h}_{QQ}^{(2)}\right]^*. \tag{7.58}
$$

The number of independent parameters are $\frac{P(P+1)}{2} + \frac{Q(Q+1)}{2}$.

Class CII: Here,

$$\mathbf{U_T} = \begin{pmatrix} \mathbf{J}_{pp} & \mathbf{0}_{pq} \\ \mathbf{0}_{qp} & \mathbf{J}_{qq} \end{pmatrix}, \quad \mathbf{U_C} = \begin{pmatrix} -\mathbf{J}_{pp} & \mathbf{0}_{pq} \\ \mathbf{0}_{qp} & \mathbf{J}_{qq} \end{pmatrix}, \quad \mathbf{U_S} = \mathbf{1}_{p,q} . \tag{7.59}$$

Given sublattice symmetry, we have the structure of the Hamiltonian as shown in Eq. (7.57). We also have $p = 2r$ and $q = 2s$. We label the p states as α_p and β_p both of which run from $1 \ldots r$. Similarly the q states can be labeled with α_q and β_q, each running from $1 \ldots s$. Transformations of these new flavors under symmetry are

$$\begin{aligned} \mathscr{T}\psi_{i\alpha_p}\mathscr{T}^{-1} &= -\psi_{i\beta_p} & \mathscr{T}\psi_{i\beta_p}\mathscr{T}^{-1} &= \psi_{i\alpha_p} \\ \mathscr{T}\psi_{i\alpha_q}\mathscr{T}^{-1} &= -\psi_{i\beta_q} & \mathscr{T}\psi_{i\beta_q}\mathscr{T}^{-1} &= \psi_{i\alpha_q}. \end{aligned} \tag{7.60}$$

The $P = pq = 4rs$ states can be rearranged as

$$\begin{pmatrix} rs \left\{ \psi_{i\alpha_p}\psi_{j\alpha_q} \right. \\ rs \left\{ \psi_{i\alpha_p}\psi_{j\beta_q} \right. \\ rs \left\{ \psi_{i\beta_p}\psi_{j\alpha_q} \right. \\ rs \left\{ \psi_{i\beta_p}\psi_{j\beta_q} \right. \end{pmatrix} \longrightarrow \begin{pmatrix} \sigma \begin{cases} rs \left\{ \psi_{i\alpha_p}\psi_{j\alpha_q} + \psi_{i\beta_p}\psi_{j\beta_q} \right. \\ rs \left\{ \psi_{i\alpha_p}\psi_{j\beta_q} - \psi_{i\beta_p}\psi_{j\alpha_q} \right. \end{cases} \\ \pi \begin{cases} rs \left\{ \psi_{i\alpha_p}\psi_{j\alpha_q} - \psi_{i\beta_p}\psi_{j\beta_q} \right. \\ rs \left\{ \psi_{i\alpha_p}\psi_{j\beta_q} + \psi_{i\beta_p}\psi_{j\alpha_q} \right. \end{cases} \end{pmatrix}. \tag{7.61}$$

Remembering that $Q = \binom{p}{2} + \binom{q}{2}$, both the $\binom{p}{2}$ and $\binom{q}{2}$ states can also be organized into symmetric and antisymmetric states. For example rearrangement of the $\binom{p}{2}$ states leads to

$$\begin{pmatrix} \binom{r}{2} \left\{ \psi_{i\alpha_p}\psi_{j\alpha_p} \right. \\ r^2 \left\{ \psi_{i\alpha_p}\psi_{j\beta_p} \right. \\ \binom{r}{2} \left\{ \psi_{i\beta_p}\psi_{j\beta_p} \right. \end{pmatrix} \longrightarrow \begin{pmatrix} \sigma \begin{cases} \binom{r}{2} \left\{ \psi_{i\alpha_p}\psi_{j\alpha_p} + \psi_{i\beta_p}\psi_{j\beta_p} \right. \\ r \left\{ \psi_{i\alpha_p}\psi_{i\beta_p} \right. \\ \binom{r}{2} \left\{ \psi_{i\alpha_p}\psi_{j\beta_p} + \psi_{j\alpha_p}\psi_{i\beta_p} \; (i \neq j) \right. \end{cases} \\ \pi \begin{cases} \binom{r}{2} \left\{ \psi_{i\alpha_p}\psi_{j\beta_p} - \psi_{j\alpha_p}\psi_{i\beta_p} \; (i \neq j) \right. \\ \binom{r}{2} \left\{ \psi_{i\alpha_p}\psi_{j\alpha_p} - \psi_{i\beta_p}\psi_{j\beta_p} \right. \end{cases} \end{pmatrix}, \tag{7.62}$$

which transform as

$$\mathscr{T}\sigma\mathscr{T}^{-1} = \sigma \quad \mathscr{T}\pi\mathscr{T}^{-1} = -\pi. \tag{7.63}$$

It is interesting to note that this rearrangement of the states do not mix the P and Q blocks.

The two-body $\mathbf{U_T}^{(2)}$ can now be written as

$$\mathbf{U_T}^{(2)} = \begin{pmatrix} \mathbf{1}_{A,B} & \mathbf{0} \\ \mathbf{0} & \mathbf{1}_{C,D} \end{pmatrix} \tag{7.64}$$

where $A = 2rs$ and $B = 2rs$, $C = r^2 + s^2$ and $D = 2\binom{r}{2} + 2\binom{s}{2}$. Imposing symmetry demands

$$\mathscr{T}\boldsymbol{\Psi}^\dagger\boldsymbol{\Psi}^\dagger\mathbf{H}^{(2)}\boldsymbol{\Psi}\boldsymbol{\Psi}\mathscr{T}^{-1} = \boldsymbol{\Psi}^\dagger\boldsymbol{\Psi}^\dagger\mathbf{H}^{(2)}\boldsymbol{\Psi}\boldsymbol{\Psi}$$
$$\mathbf{U}_\mathrm{T}^{(2)}\left[\mathbf{H}^{(2)}\right]^*\mathbf{U}_\mathrm{T}^{(2)\dagger} = \mathbf{H}^{(2)}, \tag{7.65}$$

and $\mathbf{h}_{PP}^{(2)}$, $\mathbf{h}_{QQ}^{(2)}$ remain decoupled. Writing

$$\mathbf{h}_{PP}^{(2)} = \begin{pmatrix} \mathbf{h}_{AA}^{(2)} & \mathbf{h}_{AB}^{(2)} \\ \left[\mathbf{h}_{AB}^{(2)}\right]^\dagger & \mathbf{h}_{BB}^{(2)} \end{pmatrix} \tag{7.66}$$

and implementing the above conditions, we get the following constraints

$$\mathbf{h}_{AA}^{(2)} = \left[\mathbf{h}_{AA}^{(2)}\right]^*, \quad \mathbf{h}_{BB}^{(2)} = \left[\mathbf{h}_{BB}^{(2)}\right]^*, \quad \mathbf{h}_{AB}^{(2)} = -\left[\mathbf{h}_{AB}^{(2)}\right]^*. \tag{7.67}$$

This leads to total $\frac{A(A+1)}{2} + \frac{B(B+1)}{2} + AB$ independent parameters. Similarly setting

$$\mathbf{h}_{QQ}^{(2)} = \begin{pmatrix} \mathbf{h}_{CC}^{(2)} & \mathbf{h}_{CD}^{(2)} \\ \left[\mathbf{h}_{CD}^{(2)}\right]^\dagger & \mathbf{h}_{DD}^{(2)} \end{pmatrix} \tag{7.68}$$

and implementing the symmetry conditions, we get the following constraints

$$\mathbf{h}_{CC}^{(2)} = \left[\mathbf{h}_{CC}^{(2)}\right]^*, \quad \mathbf{h}_{DD}^{(2)} = \left[\mathbf{h}_{DD}^{(2)}\right]^*, \quad \mathbf{h}_{CD}^{(2)} = -\left[\mathbf{h}_{CD}^{(2)}\right]^*, \tag{7.69}$$

leading to $\frac{C(C+1)}{2} + \frac{D(D+1)}{2} + CD$ independent parameters for this part of the Hamiltonian.

Class CI: This class arises with

$$\mathbf{U}_\mathrm{T} = \mathbf{F}, \quad \mathbf{U}_\mathrm{C} = -\mathbf{J}, \quad \mathbf{U}_\mathrm{S} = \mathbf{1}_{p,q}. \tag{7.70}$$

Sublattice symmetry already implies a structure of Hamiltonian as given in Eq. (7.57). Also we have $P = M^2$ and $Q = 2\binom{M}{2}$. States can be re-organized as

$$\begin{pmatrix} M^2 \left\{\psi_{i\alpha}\psi_{j\beta} \right. \\ \binom{M}{2}\left\{\psi_{i\alpha}\psi_{j\alpha} \right. \\ \binom{M}{2}\left\{\psi_{i\beta}\psi_{j\beta} \right. \end{pmatrix} \longrightarrow \begin{pmatrix} \sigma\left\{\binom{M}{2}\left\{\psi_{i\alpha}\psi_{j\beta} - \psi_{j\alpha}\psi_{i\beta} \;_{(i\neq j)} \right.\right. \\ \pi\left\{\binom{M}{2}\left\{\psi_{i\alpha}\psi_{j\beta} + \psi_{j\alpha}\psi_{i\beta} \;_{(i\neq j)} \right.\right. \\ M\left\{\psi_{i\alpha}\psi_{i\beta} \right. \\ \sigma\left\{\binom{M}{2}\left\{\psi_{i\alpha}\psi_{j\alpha} + \psi_{i\beta}\psi_{j\beta} \right.\right. \\ \pi\left\{\binom{M}{2}\left\{\psi_{i\alpha}\psi_{j\alpha} - \psi_{i\beta}\psi_{j\beta} \right.\right. \end{pmatrix}. \tag{7.71}$$

Again both the sectors get decoupled and are symmetric and antisymmetric under time reversal. The remaining analysis follows closely the one for class CII with the appropriate replacements of $A = \binom{M}{2}$, $B = \binom{M}{2} + M$, $C = \binom{M}{2}$ and $D = \binom{M}{2}$.

Class DIII: This one possesses

$$\mathbf{U_T} = \mathbf{J}, \quad \mathbf{U_C} = \mathbf{F}, \quad \mathbf{U_S} = \mathbf{1}_{p,q} , \tag{7.72}$$

and $L = 2M$. States can again be reorganized as symmetric and antisymmetric under time reversal as follows

$$\begin{pmatrix} M^2 \left\{ \psi_{i\alpha}\psi_{j\beta} \right. \\ \binom{M}{2} \left\{ \psi_{i\alpha}\psi_{j\alpha} \right. \\ \binom{M}{2} \left\{ \psi_{i\beta}\psi_{j\beta} \right. \end{pmatrix} \longrightarrow \begin{pmatrix} \sigma \begin{cases} \binom{M}{2} \left\{ \psi_{i\alpha}\psi_{j\beta} + \psi_{j\alpha}\psi_{i\beta} \right. & (i \neq j) \\ M \left\{ \psi_{i\alpha}\psi_{i\beta} \right. \end{cases} \\ \pi \left\{ \binom{M}{2} \left\{ \psi_{i\alpha}\psi_{j\beta} - \psi_{j\alpha}\psi_{i\beta} \right. \quad (i \neq j) \right. \\ \sigma \left\{ \binom{M}{2} \left\{ \psi_{i\alpha}\psi_{j\alpha} + \psi_{i\beta}\psi_{j\beta} \right. \right. \\ \pi \left\{ \binom{M}{2} \left\{ \psi_{i\alpha}\psi_{j\alpha} - \psi_{i\beta}\psi_{j\beta} \right. \right. \end{pmatrix} . \tag{7.73}$$

Yet again both the sectors get decoupled and are symmetric and antisymmetric under the time reversal. Rest of the analysis proceeds along the same lines as class CII and CI with $A = \binom{M}{2} + M$, $B = \binom{M}{2}$, $C = \binom{M}{2}$ and $D = \binom{M}{2}$.

7.4   *N*-body Interacting Hamiltonians

We now determine the structure of a N-body Hamiltonian of the generic form (see Eq. (7.2)),

$$\mathscr{H}^{(N)} = \sum_{\substack{i_1,i_2...,i_N \\ j_1,j_2...j_N}} (\psi_{i_1}\psi_{i_2} \cdots \psi_{i_N})^\dagger H^{(N)}_{i_1,i_2...,i_N;j_1,j_2...j_N} \psi_{j_1}\psi_{j_2} \cdots \psi_{j_N}. \tag{7.74}$$

As discussed in Sect. 7.2, this is the key ingredient to determine the structure of **H** in Eq. (7.6). A basis of $\binom{L}{N}$ states is required to describe $\mathscr{H}^{(N)}$. We find that structure determination of $\mathscr{H}^{(N)}$ depends on whether N is odd or even. In the next subsection we focus on the cases when N is odd followed by the subsection for N even. In both cases, the strategy is to use Table 2.1 to find a canonical representation of a symmetry operation $\mathbf{U}^{(N)}$, which then aids in finding the final structure of $\mathbf{H}^{(N)}$.

7.4.1   Structure of $\mathbf{H}^{(N)}$ for N Odd

Class A: This class has no symmetries and the only condition imposed on the Hamiltonian is

$$\mathbf{H}^{(N)} = \left[\mathbf{H}^{(N)}\right]^{\dagger}. \tag{7.75}$$

The Hamiltonian has $\left\{\binom{L}{N}\right\}^2$ independent parameters.

Class AI: The implementation of the time reversal demands

$$\mathbf{H}^{(N)} = \left[\mathbf{H}^{(N)}\right]^{*}. \tag{7.76}$$

Hence the number of independent parameters are $\frac{1}{2}\binom{L}{N}\left[\binom{L}{N} + 1\right]$.

Class AII: Given the conditions, as in (7.30), we know $L = 2M$. There are M α and M β states. A generic many-body state has the form

$$\alpha_a \beta_{N-a} \equiv \underbrace{\psi_{i\alpha_1} \dots \psi_{i\alpha_a}}_{a} \underbrace{\psi_{j\beta_1} \dots \psi_{j\beta_{N-a}}}_{N-a}. \tag{7.77}$$

This state transforms in the following way,

$$\mathscr{T}\alpha_a\beta_{N-a}\mathscr{T}^{-1} = \beta_a\alpha_{N-a}(-1)^a = \alpha_{N-a}\beta_a(-1)^{a(N-a+1)} = (-1)^a\alpha_{N-a}\beta_a. \tag{7.78}$$

The last equality uses the fact that N is odd. Using this we form two kinds of states, where one is made of *even* number of αs($\equiv E_\alpha$) and the other with *odd* ($\equiv O_\alpha$), having

$$\dim O_\alpha = \sum_{a=1,3\dots}^{N} \binom{M}{a}\binom{M}{N-a} = \frac{1}{2}\binom{2M}{N}$$

$$\dim E_\alpha = \sum_{a=0,2\dots}^{N-1} \binom{M}{a}\binom{M}{N-a} = \frac{1}{2}\binom{2M}{N}. \tag{7.79}$$

These new states transform conveniently as

$$\mathscr{T}O_\alpha\mathscr{T}^{-1} = -E_\alpha \qquad \mathscr{T}E_\alpha\mathscr{T}^{-1} = O_\alpha \tag{7.80}$$

which overall gives the following transformation

$$\mathscr{T}\Psi \dots \Psi\mathscr{T}^{-1} = -\mathbf{J}^{(N)}\Psi \dots \Psi \tag{7.81}$$

where $\mathbf{J}^{(N)}$ is the N-body version of $\mathbf{J}$ (defined in (2.44)), but now with dimensions $\binom{L}{N}$. This tells us that the Hamiltonian transforms as the one-body case discussed near Eq. (2.80). However, the dimension now is given by, dim $\mathcal{H}_{\text{All}}^{(N)} = P^2 + 2 \times \frac{P(P-1)}{2}$, where $P = \frac{1}{2}\binom{2M}{N}$.

Class D: In this class the transformation of ψs are shown in Eq. (7.40). The Hamiltonian satisfies

$$: \mathscr{C}\mathcal{H}^{(N)}\mathscr{C}^{-1} := \mathcal{H}^{(N)}. \tag{7.82}$$

This implies, $-[\mathbf{H}^{(N)}]^* = \mathbf{H}^{(N)}$, with the number of independent parameters being $\frac{P(P-1)}{2}$, where $P = \binom{L}{N}$.

Class C: Using the transformations for the fermionic operators given in Eq. (7.43), we find out that O_α and E_α transform as

$$\mathscr{C}O_\alpha\mathscr{C}^{-1} = (-1)^{\frac{N+1}{2}}(E_\alpha)^\dagger \qquad \mathscr{C}E_\alpha\mathscr{C}^{-1} = (-1)^{\frac{N-1}{2}}(O_\alpha)^\dagger. \tag{7.83}$$

Therefore as discussed near Eq. (2.81), we again have $\mathbf{U}_C^{(N)} \propto \mathbf{J}^{(N)}$. Hence the Hamiltonian satisfies the same condition as the single-body case, with the dimension dim $\mathcal{H}_C^{(N)} = P^2 + 2 \times \frac{P(P+1)}{2} = P(2P+1)$ where $P = \frac{1}{2}\binom{2M}{N}$.

Class AIII: Given the conditions for Class AIII (Eq. (7.50)) where p and q states are labeled by α and β, the states transform according to

$$\mathscr{S}O_\alpha\mathscr{S}^{-1} = (-1)^{\frac{N-1}{2}}(O_\alpha)^\dagger \qquad \mathscr{S}E_\alpha\mathscr{S}^{-1} = (-1)^{\frac{N+1}{2}}(E_\alpha)^\dagger, \tag{7.84}$$

having

$$\text{dim } O_\alpha = \sum_{a=1,3...}^{N} \binom{p}{a}\binom{q}{N-a} = P \qquad \text{dim } E_\alpha = \sum_{a=0,2...}^{N-1} \binom{p}{a}\binom{q}{N-a} = Q. \tag{7.85}$$

Therefore $\mathbf{U}_S^{(N)} \propto \mathbf{1}_{P,Q}$ as we had seen in the one-body case (see near Eq. (2.82)). This implies that just like the one-body case, the diagonal blocks of $\mathbf{H}^{(N)}$ will be constrained to be *zero* (see Eq. (2.83)) forcing the structure of the Hamiltonian to be

$$\mathbf{H}^{(N)} = \begin{pmatrix} \mathbf{0}_{PP} & \mathbf{h}_{PQ}^{(N)} \\ \left[\mathbf{h}_{PQ}^{(N)}\right]^\dagger & \mathbf{0}_{QQ} \end{pmatrix}. \tag{7.86}$$

This gives the number of independent parameters as dim $\mathcal{H}_{\text{AIII}}^{(N)} = 2PQ$.

We again bring to attention of the reader that from here onwards and until the end of this subsection all the classes have sublattice symmetry and therefore their respective Hamiltonians will always posses the above form.

Class BDI: The time reversal symmetry implementation on the class AIII, demands $\mathbf{H}^{(N)}$ to be real (see Eq. (7.86)) and therefore, the dim $\mathcal{H}^{(N)}_{\mathsf{BDI}} = PQ$.

Class CII: The transformation of states is given by Eq. (7.60), and in this class we have $p = 2r, q = 2s$. We now show how these affect the P and Q states. The general structure of any state is of the form

$$\underbrace{\underbrace{\psi_{i\alpha_{p_1}} \cdots \psi_{i\alpha_{p_c}}}_{c} \underbrace{\psi_{i\beta_{p_1}} \cdots \psi_{i\beta_{p_{a-c}}}}_{a-c}}_{a} \underbrace{\underbrace{\psi_{j\alpha_{q_1}} \cdots \psi_{j\alpha_{q_d}}}_{d} \underbrace{\psi_{j\beta_{q_1}} \cdots \psi_{j\beta_{q_{N-a-d}}}}_{N-a-d}}_{N-a} . \qquad (7.87)$$

Given sublattice symmetry and the fact that $\alpha_p \leftrightarrow \beta_p$, $\alpha_q \leftrightarrow \beta_q$ under time reversal, the transformed states still live within their respective blocks. Therefore they do not change the constraints put on the Hamiltonian due to sublattice symmetry and hence preserve the structure of the Hamiltonian appearing in Eq. (7.86).

  To see this more clearly, we write the above state schematically and find that it transforms under $\mathcal{T}$ as

$$\alpha_{p_c}\beta_{p_{a-c}}\alpha_{q_d}\beta_{q_{N-a-d}} \xrightarrow{\mathcal{T}} \begin{cases} (-1)^c \alpha_{p_{a-c}}\beta_{p_a}\alpha_{q_{N-a-d}}\beta_{q_d} & ; a \text{ odd} \\ (-1)^d \alpha_{p_{a-c}}\beta_{p_a}\alpha_{q_{N-a-d}}\beta_{q_d} & ; a \text{ even} \end{cases} . \qquad (7.88)$$

Therefore the odd-even structure of a is still preserved. On further substructuring of states into odd(even)-number of $\alpha_p \equiv O_{\alpha_p}(E_{\alpha_p})$ and odd(even)-number of $\alpha_q \equiv O_{\alpha_q}(E_{\alpha_q})$ within each P and Q block, we find that $\mathcal{T}$ symmetry acts on them as

$$\begin{pmatrix} P\begin{cases}O_\alpha \\ \\ Q\begin{cases}E_\alpha\end{cases}\end{cases} \end{pmatrix} \longrightarrow \begin{pmatrix} 1\begin{cases}\frac{P}{2}\end{cases}\begin{bmatrix}E_{\alpha_p}O_{\alpha_q}\\ E_{\alpha_p}E_{\alpha_q}\end{bmatrix} \\ 2\begin{cases}\frac{P}{2}\end{cases}\begin{bmatrix}O_{\alpha_p}O_{\alpha_q}\\ O_{\alpha_p}E_{\alpha_q}\end{bmatrix} \\ 3\begin{cases}\frac{Q}{2}\end{cases}\begin{bmatrix}O_{\alpha_p}E_{\alpha_q}\\ E_{\alpha_p}E_{\alpha_q}\end{bmatrix} \\ 4\begin{cases}\frac{Q}{2}\end{cases}\begin{bmatrix}O_{\alpha_p}O_{\alpha_q}\\ E_{\alpha_p}O_{\alpha_q}\end{bmatrix} \end{pmatrix} \xrightarrow{\mathcal{T}} \begin{pmatrix} 2\begin{cases}O_{\alpha_p}O_{\alpha_q}\\ O_{\alpha_p}E_{\alpha_q}\end{cases} \\ -1\begin{cases}-E_{\alpha_p}O_{\alpha_q}\\ -E_{\alpha_p}E_{\alpha_q}\end{cases} \\ 4\begin{cases}O_{\alpha_p}O_{\alpha_q}\\ E_{\alpha_p}O_{\alpha_q}\end{cases} \\ -3\begin{cases}-O_{\alpha_p}E_{\alpha_q}\\ -E_{\alpha_p}E_{\alpha_q}\end{cases} \end{pmatrix} . \qquad (7.89)$$

Hence $\mathbf{U}_T^{(N)}$ takes the same form as in the one-body case (see Eq. (2.69)) which is

$$\mathbf{U}_T^{(N)} = \begin{pmatrix} \mathbf{J}_{PP} & \mathbf{0}_{PQ} \\ \mathbf{0}_{QP} & \mathbf{J}_{QQ} \end{pmatrix} . \qquad (7.90)$$

The structure of Hamiltonian then follows from the one-body case discussed near Eq. (2.85) with the number of independent parameters given by dim $\mathcal{H}^{(N)}_{\mathsf{CII}} = PQ$.

Class CI: The presence of sublattice symmetry enforces that the Hamiltonian has a off-diagonal structure (see Eq. (7.86)) with dimension $P = Q = \frac{1}{2}\binom{2M}{N}$. As we have seen, $\mathbf{U_T}$ in this case is $\mathbf{F}$ (see Eq. (2.73)), and converts $\alpha \leftrightarrow \beta$ and vice-versa, therefore the transformation is

$$\mathcal{T} O_\alpha \mathcal{T}^{-1} = E_\alpha \qquad \mathcal{T} E_\alpha \mathcal{T}^{-1} = O_\alpha. \tag{7.91}$$

Hence $\mathbf{U_T^{(N)}}$ is $\mathbf{F}^{(N)}$ (a generalized version of $\mathbf{F}$ defined in (2.72) but with dimension $\binom{L}{N}$) and the constraint on $\mathbf{H}^{(N)}$ is same as that shown in Eq. (2.89) but with M replaced with P. The dim of $\mathcal{H}_{CI}^{(N)}$ is $P(P+1)$.

Class DIII: For this class, $\mathbf{U_T} = \mathbf{J}$ (see Eq. (2.74)) and states transform as

$$\mathcal{T} \psi_{i\alpha} \mathcal{T}^{-1} = -\psi_{i\beta} \qquad \mathcal{T} \psi_{i\beta} \mathcal{T}^{-1} = \psi_{i\alpha}. \tag{7.92}$$

The odd and even α states transform in the following way

$$\mathcal{T} O_\alpha \mathcal{T}^{-1} = -E_\alpha \qquad \mathcal{T} E_\alpha \mathcal{T}^{-1} = O_\alpha. \tag{7.93}$$

Therefore $\mathbf{U_T^{(N)}} = \mathbf{J}^{(N)}$ with matrix dimension P. The constraints on the Hamiltonian is same as that shown in Eq. (2.90) but now with dim $\mathcal{H}_{DIII}^{(N)} = P(P-1)$, where $P = \frac{1}{2}\binom{2M}{N}$. The generic structure of $\mathbf{H}^{(N)}$ for N-odd is summarized and tabulated in Table 7.2.

7.4.2 Structure of $\mathbf{H}^{(N)}$ for N Even

Class A: This class has no symmetries and the only condition that applies is $\mathbf{H}^{(N)} = \left[\mathbf{H}^{(N)}\right]^\dagger$. The Hamiltonian therefore has $\left\{\binom{L}{N}\right\}^2$ independent parameters.

Class AI: As is in the case of N-odd, this class only enforces the condition $\mathbf{H}^{(N)} = \left[\mathbf{H}^{(N)}\right]^*$ and the number of independent parameters are $\frac{1}{2}\binom{L}{N}\left\{\binom{L}{N}+1\right\}$.

Class AII: Given the conditions in Eq. (7.30), we know $L = 2M$. There are M α and M β states. Note that given N is even, we can reorganize states into symmetric and antisymmetric states (like in two-body case, see Eq. (7.33)). The transformation on a generic state (see Eq. (7.77)) is given by

$$\mathcal{T} \alpha_a \beta_{N-a} \mathcal{T}^{-1} = \alpha_{N-a} \beta_a. \tag{7.94}$$

Therefore symmetric and antisymmetric states can be made by linearly combining, $\alpha_a \beta_{N-a}$ and $\alpha_{N-a}\beta_a$ with a $\pm$ sign. Note that π and σ don't contain the same number of states. The state $\alpha_{\frac{N}{2}} \beta_{\frac{N}{2}}$ (with same orbital labels) goes back to itself without any sign change under transformation and therefore is a symmetric state. Then dimensions of symmetric(σ) and antisymmetric(π) states are

$$\dim \sigma = \frac{1}{2}\left\{\binom{L}{N} + \binom{M}{N/2}\right\} = P \qquad \dim \pi = \frac{1}{2}\left\{\binom{L}{N} - \binom{M}{N/2}\right\} = Q$$
(7.95)

and $\mathbf{U}_T^{(N)} = \mathbf{1}_{P,Q}$. The structure of the Hamiltonian in this basis is given by

$$\mathbf{H}^{(N)} = \begin{pmatrix} \mathbf{h}_{PP}^{(N)} & \mathbf{h}_{PQ}^{(N)} \\ \left[\mathbf{h}_{PQ}^{(N)}\right]^\dagger & \mathbf{h}_{QQ}^{(N)} \end{pmatrix}$$
(7.96)

with symmetry conditions being

$$\mathscr{T}\mathbf{\Psi}^\dagger\mathbf{\Psi}^\dagger\mathbf{H}^{(N)}\mathbf{\Psi}\mathbf{\Psi}\mathscr{T}^{-1} = \mathbf{\Psi}^\dagger\mathbf{\Psi}^\dagger\mathbf{H}^{(N)}\mathbf{\Psi}\mathbf{\Psi}$$
$$\mathbf{U}_T^{(N)}\left[\mathbf{H}^{(N)}\right]^*\mathbf{U}_T^{(N)\dagger} = \mathbf{H}^{(N)}.$$
(7.97)

This imposes the following constraints,

$$\mathbf{h}_{PP}^{(N)} = \left[\mathbf{h}_{PP}^{(N)}\right]^*, \quad \mathbf{h}_{QQ}^{(N)} = \left[\mathbf{h}_{QQ}^{(N)}\right]^*, \quad \mathbf{h}_{PQ}^{(N)} = -\left[\mathbf{h}_{PQ}^{(N)}\right]^*.$$
(7.98)

The total number of independent parameters are $\frac{P(P+1)}{2} + \frac{Q(Q+1)}{2} + PQ$. This general formula reduces to the specific two-body case discussed above by substituting $N = 2$.

Class D: For this class the transformation of ψs are determined by Eq. (7.40). The constraint on the Hamiltonian is then $\left[\mathbf{H}^{(N)}\right]^* = \mathbf{H}^{(N)}$, which is same as that for Class AI.

Class C: The symmetric (σ) and antisymmetric states (π) are again linear combinations of $\alpha_a\beta_{N-a}$ and $\alpha_{N-a}\beta_a$ with a $\pm$ sign. However depending on the value of N, the $+$ linear combination may transform under the $\mathscr{C}$ symmetry with a $-$ sign and therefore be an antisymmetric state by definition. In general a state will transform as

$$\alpha_a\beta_{N-a} \xrightarrow{\mathscr{C}} (-1)^{N/2}(\alpha_{N-a}\beta_a)^\dagger.$$
(7.99)

With the definition

$$P = \frac{1}{2}\left\{\binom{L}{N} + \binom{M}{N/2}\right\} \qquad Q = \frac{1}{2}\left\{\binom{L}{N} - \binom{M}{N/2}\right\},$$
(7.100)

the dimensions of symmetric and antisymmetric states now are

$$\dim \sigma = \begin{cases} P & ; N/2 \text{ even} \\ Q & ; \text{ odd} \end{cases} \qquad \dim \pi = \begin{cases} Q & ; N/2 \text{ even} \\ P & ; \text{ odd} \end{cases}.$$
(7.101)

Therefore $\mathbf{U}_C^{(N)} = (-1)^{N/2}\mathbf{1}_{P,Q}$. The structure of the Hamiltonian is again constrained in the same way as we had seen in the Eqs. (7.96) and (7.98).

Class AIII: Given the conditions for class AIII (Eq. (7.50)) where p and q states are labeled by α and β, we look at the transformation of E_α and O_α states introduced near Eq. (7.77),

$$\mathscr{S}O_\alpha\mathscr{S}^{-1} = (-1)^{\frac{N}{2}-1}O_\alpha^\dagger \qquad \mathscr{S}E_\alpha\mathscr{S}^{-1} = (-1)^{\frac{N}{2}}E_\alpha^\dagger. \qquad (7.102)$$

The dimensions are given by,

$$\dim O_\alpha = \sum_{a=1,3,\ldots}^{N-1} \binom{p}{a}\binom{q}{N-a} = P \qquad \dim E_\alpha = \sum_{a=0,2,\ldots}^{N} \binom{p}{a}\binom{q}{N-a} = Q, \qquad (7.103)$$

and $\mathbf{U}_S^{(N)} = (-1)^{\frac{N}{2}-1}\mathbf{1}_{P,Q}$. This imposes the condition that off-diagonal blocks in Eq. (7.96) are now *zero* and the independent parameters, which are $P^2 + Q^2$ in number, belong to the diagonal blocks.

The following classes discussed in this subsection all have sublattice symmetry and the form of the Hamiltonian in each class will satisfy

$$\mathbf{H}^{(N)} = \begin{pmatrix} \mathbf{h}_{PP}^{(N)} & \mathbf{0}_{PQ} \\ \mathbf{0}_{QP} & \mathbf{h}_{QQ}^{(N)} \end{pmatrix} \qquad (7.104)$$

Class BDI: Apart from the implementation of the sublattice symmetry as in the previous section, time reversal operation for this class is just $\mathbf{U}_T^{(N)} = \mathbf{1}$ and therefore just demands $\mathbf{h}_{PP}^{(N)}$ and $\mathbf{h}_{QQ}^{(N)}$ to be real. The number of independent parameters are $\frac{P(P+1)}{2} + \frac{Q(Q+1)}{2}$.

Class CII: This class requires that $p = 2r$ and $q = 2s$. Imposition of sublattice symmetry breaks the basis into blocks O_α and E_α of dimensions given in Eq. (7.103). Now the α states are made of two varieties α_p and β_p. While the β states are made of α_q and β_q. The time reversal symmetry converts $\alpha_p \leftrightarrow \beta_p$ and $\alpha_q \leftrightarrow \beta_q$ states. Therefore one can make symmetric and antisymmetric combinations. The transformations under $\mathscr{T}$ are given in Eq. (7.60). Now let us discuss the O_α states. Dimension of O_α states is P and is comprised of following kinds of states

$$\left(\begin{array}{l} 1 \begin{cases} E_{\alpha_p}E_{\alpha_q} \\ O_{\alpha_p}O_{\alpha_q} \end{cases} \\ 2 \begin{cases} O_{\alpha_p}E_{\alpha_q} \\ E_{\alpha_p}O_{\alpha_q} \end{cases} \end{array} \right) \xrightarrow{\mathscr{T}} \left(\begin{array}{l} 1 \begin{cases} O_{\alpha_p}O_{\alpha_q} \\ E_{\alpha_p}E_{\alpha_q} \end{cases} \\ 2 \begin{cases} -E_{\alpha_p}O_{\alpha_q} \\ -O_{\alpha_p}E_{\alpha_q} \end{cases} \end{array} \right). \qquad (7.105)$$

Hence the number of symmetric states($\equiv A$) and number of antisymmetric states($\equiv B$) equals $P/2$ within the P block (see Table 7.1).

A typical E_α state transforms under $\mathscr{T}$ as

$$\alpha_{p_c}\beta_{p_{a-c}}\alpha_{q_d}\beta_{q_{N-a-d}} \xrightarrow{\mathscr{T}} \alpha_{p_{a-c}}\beta_{p_c}\alpha_{q_{N-a-d}}\beta_{q_d}, \tag{7.106}$$

using which symmetric(σ) and antisymmetric(π) combinations can again be formed. For each a, the total number of states ($\sigma + \pi$) are,

$$\binom{p}{a}\binom{q}{N-a} = \sum_{c,d}\binom{r}{c}\binom{r}{a-c}\binom{s}{d}\binom{s}{N-a-d}. \tag{7.107}$$

The number of symmetric and antisymmetric states are equal except when $c = \frac{a}{2}$ and $d = \frac{N-a}{2}$ which gives individual counts of symmetric and antisymmetric states as

$$\sigma_a = \frac{1}{2}\left\{\binom{p}{a}\binom{q}{N-a} + \binom{r}{\frac{a}{2}}\binom{s}{\frac{N-a}{2}}\right\}$$

$$\pi_a = \frac{1}{2}\left\{\binom{p}{a}\binom{q}{N-a} - \binom{r}{\frac{a}{2}}\binom{s}{\frac{N-a}{2}}\right\}. \tag{7.108}$$

This brings the total symmetric (C) and antisymmetric (D) states in the Q block to be

$$C = \sum_{a=0,2,\ldots}^{N} \sigma_a \qquad D = \sum_{a=0,2,\ldots}^{N} \pi_a. \tag{7.109}$$

The structure of the Hamiltonian for the O_α states is similar to the All case with (P, Q) of All replaced with (A, B) respectively. Likewise, replacing (P, Q) with (C, D) gives us the Hamiltonian structure for E_α states(see Table 7.1). The number of independent parameters are

$$\frac{A(A+1)}{2} + \frac{B(B+1)}{2} + AB + \frac{C(C+1)}{2} + \frac{D(D+1)}{2} + CD. \tag{7.110}$$

Class CI: In this class, $L = 2M$. Imposition of sublattice symmetry again breaks the basis states into two blocks P and Q with dimensions as seen in Eq. (7.103). It can be seen that $\alpha_a\beta_{N-a} \to (-1)^a\alpha_{N-a}\beta_a$. A state having a as odd(even) belongs to the $P(Q)$ block. Therefore symmetric and antisymmetric states can again be constructed. Given $\frac{N}{2}$ is an even integer, then the number of symmetric and antisymmetric states which can be formed using states of the $P(Q)$ block will be equal(unequal). This makes the P, Q Hamiltonian blocks take the structure of the Hamiltonian for the case CII with

$$A = \frac{P}{2} \qquad C = \frac{1}{2}\left\{Q + \begin{pmatrix} M \\ N/2 \end{pmatrix}\right\}$$

$$B = \frac{P}{2} \qquad D = \frac{1}{2}\left\{Q - \begin{pmatrix} M \\ N/2 \end{pmatrix}\right\}. \tag{7.111}$$

Where as, for $N/2$ odd the number of symmetric and antisymmetric states become unequal(equal) leading to the same Hamiltonian structure(also see Table 7.1) as before but with

$$A = \frac{1}{2}\left\{P - \begin{pmatrix} M \\ N/2 \end{pmatrix}\right\} \qquad C = \frac{Q}{2}$$

$$B = \frac{1}{2}\left\{P + \begin{pmatrix} M \\ N/2 \end{pmatrix}\right\} \qquad D = \frac{Q}{2}. \tag{7.112}$$

The expression for total number of independent parameters is same as Class CII.

Class DIII: The reasoning for DIII class is similar to CI with the crucial distinction that $\alpha_a\beta_{N-a} \to \alpha_{N-a}\beta_a$. This distinction readjusts the number of symmetric and antisymmetric states in P, Q blocks. The similarities between the two classes gives us the same cases depending on the oddness and evenness of $N/2$, however now with minor changes in the formulae for dimensions of the $A - D$ sub-blocks of P, Q, which now become

$$A = \begin{cases} P/2 \\ \frac{P}{2} + \frac{1}{2}\binom{M}{N/2} \end{cases} \qquad B = \begin{cases} P/2 & ; N/2 \text{ even} \\ \frac{P}{2} - \frac{1}{2}\binom{M}{N/2} & ; \text{ odd} \end{cases}$$

$$C = \begin{cases} \frac{Q}{2} + \frac{1}{2}\binom{M}{N/2} \\ Q/2 \end{cases} \qquad D = \begin{cases} \frac{Q}{2} - \frac{1}{2}\binom{M}{N/2} & ; N/2 \text{ even} \\ Q/2 & ; \text{ odd} \end{cases}. \tag{7.113}$$

The structure of Hamiltonian remains same as discussed for class CI with the expression for total number of independent parameters given by Eq. (7.110). The generic structure of $\mathbf{H}^{(N)}$ for N-even is summarized and tabulated in Table 7.1.

7.5   Summary and Perspective

In this chapter we have revisited the tenfold scheme of classification of fermions with the aim of studying interacting systems. We have endeavored to provide a simple and direct approach that makes clear the underlying physical content even while not being tied to the single particle picture. Tables 7.1 and 7.2 contain the essential results on the structures of the N-body Hamiltonians in each class. A key utility of the results is the possible inference of properties of generic systems in a particular symmetry class. A case in point: the *zero* modes in class AIII. In class AIII we have seen in

Table 7.3 Many-body zero modes: Number of *zero* modes ($\equiv |P - Q|$) for a few representative cases where the $N = 1, 3, 5$ and number of single particle states are of the form $p = q + k$ where k is any integer. The terms can be derived by straightforward calculation of P, Q using expressions that are given in Table 7.2

	$p = q + 1$	$p = q + k$
$N = 1$	1	k
$N = 3$	q	$\left\|\frac{k}{6}\left(6q - (k-3)k - 2\right)\right\|$
$N = 5$	$\frac{q(q-1)}{2}$	$\left\|\frac{1}{120}k\left(-20(k-3)kq + (k-5)k((k-5)k + 10) + 60q^2 - 100q + 24\right)\right\|$

Fig. 7.1 Zero modes: Number of *zero* modes ($\equiv |P - Q|$) as a function of the order of interaction N, in class AIII for $p = q + 1$, and $q = 20, 40$ and 60. Note the initial exponential rise in the number of zero modes as the order of interaction increases

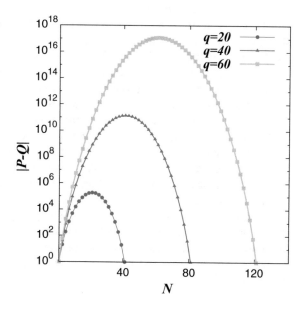

the previous sections that the many-body Hamiltonian has a off-diagonal structure when the order of interaction—N is odd (see Table 7.2). If the off-diagonal blocks of such a Hamiltonian is not of the same dimension, one has to tolerate nonnegotiable *zero* energy states, irrespective of the energetics involved in the Hamiltonian. This for example can be trivially seen in the case of noninteracting case, where if the number of single particle states p is not equal to q (see Table 2.2) then one gets $p - q$ *zero* modes. The same condition for many body Hamiltonians will lead to $P - Q$ *zero* modes (see Table 7.2). Considering a system which has $p \neq q$, the number of modes has an interesting dependence on the order of interaction N. With increasing N, there is an exponential rise in the number of *zero* modes which is proportional to $k = p - q$ (see Table 7.3). It is interesting to note the dependence of the *zero* modes with further increase in N (see Fig. 7.1).

There are several areas of contemporary research where these results can have important implications:

7.5.1   Topological Classification of Interacting Systems

Topological insulators and other topological phases for noninteracting systems form a part of a broader class of phases called the Symmetry Protected Topological (SPT) phases. Such phases are protected by symmetries and, by definition, does not have long-range entanglement, topological order or fractional excitations (for recent reviews, see [1–5]). However surface theory of such systems are still anomalous in the sense that it cannot be otherwise be realized as lattice models in the particular surface dimension. Given a spatial dimension and interactions, how many such distinct phases are possible? This question has been addressed by a number of authors (see [1–5] and references therein). Based on group cohomology and supercoholomogy, steps have been taken towards a complete classification of all interacting bosonic and fermionic phases [6]. Classification of time reversal symmetric interacting three dimensional phases has also been achieved by analyzing the monopole quantum numbers [7]. Some of these insights into the classification has been motivated by construction of explicit microscopic models [8–10]. For example, Fidkowski and Kitaev showed how a noninteracting $\mathbb{Z}$ classification breaks down to $\mathbb{Z}_8$ in a one dimensional model [9]. Haldane's spin-1 chain continues to be a paradigmatic example of a SPT phase of bosons [8]. However, such models are few and are quite nontrivially "designed". In many of these systems the implementation of the symmetry is done case-by-case basis, and it is challenging to tie them in a common framework, as is otherwise possible for the noninteracting cases. Our results can provide a systematic framework to investigate SPT phases in different symmetry classes, in that it provides a prescription to construct Hamiltonian which includes disorder etc. Another interesting point worthy of investigation is regarding the "geometry" of interacting systems in a given class. In the noninteracting case, there is a connection between the set of all Hamiltonians in each class and Cartan's symmetric spaces. What is the analogous geometric picture for interacting systems?

7.5.2   Sachdev–Ye–Kitaev Models

Sachdev and Ye devised a model [11], which contains only quartic terms (in our notation, $N = 2$) and has *all to all* random interactions. This model is given by

$$\mathcal{H} = \sum_{i,j,k,l=1}^{L} J_{ijkl} \psi_i^\dagger \psi_j^\dagger \psi_k \psi_l \qquad (7.114)$$

where J_{ijkl} are random and uncorrelated (which imbibe constraints imposed by fermion antisymmetry and hermiticity) with a mean zero and a finite variance which sets the energy scale. The model displays a strange metallic phase and has relations to two-dimensional gravity as was recently shown by Kitaev [12]—leading to the surge of interest in this model. The rich physics contained in this model has things to offer for diverse areas of physics from quantum gravity to condensed matter physics. For example, several recent works [13–16] have provided higher dimensional generalizations that have potential to throw light on equilibrium and out-of-equilibrium properties of strange metals. This active area can benefit from our results—for example, one can address the question: how does the physics of Sachdev–Ye–Kitaev model depend on its non-ordinary symmetries?

7.5.3 Random Matrix Theory and Many-Body Localization

Another recent direction in condensed matter physics has been that of many-body localization phenomena and the questions pertaining to ergodicity, thermalization etc. [17, 18]. These questions are also intimately related to random matrix theory. Random matrix theory originally devised to understand nuclear spectra has since been used to gain insights into variety of phenomena such as chaos, localization (see [19, 20]). Given a matrix of size $L \times L$ one can obtain the mean of ratio of adjacent energy gaps, it has been shown that this ratio takes a value of ~ 0.39 for many-body localized phase and ~ 0.53 for a delocalized phase [21]. Interestingly these are also the values if one takes a random matrix which satisfy Poisson or Gaussian-orthogonal ensemble (GOE) level statistics respectively. In fact, the tenfold way has deep connections to this physics. A generic Hamiltonian belonging to any symmetry class has interesting implications on its eigenvalue statistics and level spacing (see [22] and references therein). Our results open up avenues to analyze such physics for interacting systems. In fact, an interesting observation from our results is that for a generic interacting Hamiltonian in classes such as **A** and **AI** (with only N-body interactions), where every orbital talks to every other randomly, is never in a many-body localized phase. This statement is seen by the fact that for none of the generic Hamiltonians in such classes for interacting systems (see Tables 7.1 and 7.2) do we have a system which allows for Poisson level statistics [23]. An interesting theoretical direction is to study disordered interacting Hamiltonians with correlated (or banded) entries to obtain an analytical handle on the many-body localization problem. The physics of many-body localization in each symmetry class can also be investigated.

References

1. Chiu CK, Teo JCY, Schnyder AP, Ryu S (2016) Classification of topological quantum matter with symmetries. Rev Mod Phys 88:035005
2. Ludwig AWW (2016) Topological phases: classification of topological insulators and superconductors of non-interacting fermions, and beyond. Phys Scr 2016(T168):014001. arXiv:1512.08882
3. Ryu S (2015) Interacting topological phases and quantum anomalies. Phys Scr T164:014009
4. Senthil T (2015) Symmetry-protected topological phases of quantum matter. Ann Rev Condens Matter Phys 6:299–324

5. Wen X-G (2016) Zoo of quantum-topological phases of matter, pp 1–16. arXiv:1610.03911
6. Wang Q-R, Gu Z-C (2017) Towards a complete classification of fermionic symmetry protected topological phases in 3D and a general group supercohomology theory. pp 1–39
7. Wang C, Potter AC, Senthil T (2014) Classification of interacting electronic topological insulators in three dimensions. Science (New York, NY) 343(6171):629–631
8. Haldane FDM (1983) Nonlinear field theory of large-spin Heisenberg antiferromagnets: semiclassically quantized solitons of the one-dimensional easy-axis Néel state. Phys Rev Lett 50:1153–1156
9. Fidkowski L, Kitaev A (2010) Effects of interactions on the topological classification of free fermion systems. Phys Rev B 81:134509
10. Turner AM, Pollmann F, Berg E (2011) Topological phases of one-dimensional fermions: an entanglement point of view. Phys Rev B Condens Matter Mater Phys 83(7):1–11
11. Sachdev S, Ye J (1993) Gapless spin-fluid ground state in a random quantum Heisenberg magnet. Phys Rev Lett 70:3339–3342
12. Kitaev A (2015) A simple model of quantum holography
13. Gu Y, Qi X-L, Stanford D (2016) Local criticality, diffusion and chaos in generalized Sachdev-ye-Kitaev models. arXiv preprint, arXiv:1609.07832
14. Davison RA, Fu W, Georges A, Gu Y, Jensen K, Sachdev S (2016) Thermoelectric transport in disordered metals without quasiparticles: the Syk models and holography. arXiv preprint, arXiv:1612.00849
15. Banerjee S, Altman E (2016) Solvable model for a dynamical quantum phase transition from fast to slow scrambling. arXiv preprint, arXiv:1610.04619
16. Jian S-K, Yao H (2017) Solvable Syk models in higher dimensions: a new type of many-body localization transition. arXiv:1703.02051
17. Altman E, Vosk R (2015) Universal dynamics and renormalization in many-body-localized systems. Ann Rev Condens Matter Phys 6(1):383–409
18. Nandkishore R, Huse DA (2015) Many body localization and thermalization in quantum statistical mechanics. Ann Rev Condens Matter Phys 6(1):15–38
19. Mehta ML (2005) Random matrices. ISBN 9780120884094, arXiv:0509286
20. Akemann G, Baik J, Di Francesco P (2011) The Oxford handbook of random matrix theory. Oxford University Press, Oxford
21. Pal A, Huse DA (2010) Many-body localization phase transition. Phys Rev B 82:174411
22. Beenakker CWJ (2015) Random-matrix theory of majorana fermions and topological superconductors. Rev Mod Phys 87:1037–1066
23. Haake F (2013) Quantum signatures of chaos, vol 54. Springer Science & Business Media, Berlin

Chapter 8
Epilogue

A sleeper class train journey is all you need. The rushing wind, the beautiful land-scape, the amazing people this country hosts. The projectile voices, the myriad of noises and—the kids waving at you with their toothy smiles, as your train passes by their half clad homes. You could find countless "*ill*" systems here, and could also find incredible *order*s develop. On one such journey a guy interrupted me to ask "Sir, what do you do?" I responded, also with a request to address me by my name. He replied, now with some more people listening intently, "*Sir*, what is your research on?" That such materials exist, which can conduct only on the surface was new to them. The conversations don't end there. As you speak about how many birds make patterns, how glasses flow, about fluids which can climb walls, about the trains which levitate—the pupils dilate. Condensed matter surrounds us, and with a billion curiosities around, its lucky to be allowed to explore a few. In these last chapters we have visited some of these explorations. Some of the successful ones have been recounted, many others which led to dark alleys, pitfalls and bruises have been kept secret.

We started the narrative in Chap. 1 with a bird's eye view of the condensed matter landscape. We visited the ideas of phases and their transitions which are built on the Landau–Wilson–Ginzburg paradigm. We revisited the early studies in *ill* condensed matter in form of Kondo effect, scaling theory of localization etc. We next intro-duced the questions condensed matter community is worried about and motivated the problems we wish to visit in this thesis. The primary aim was to examine the recent developments in the field such as topological phases of matter, role of few intrinsic symmetries, spin-orbit coupling etc. in context of *ill* systems. Some of these aspects have been brought to light in the course of last chapters. However, many new questions have also been raised. In this concluding section we revisit our findings, discuss the questions and provide possibilities for future scope.

In Chap. 2, we posed the question "How does one write a Hamiltonian in a par-ticular symmetry class, on any *ill* motif—say a polymer?" We began by building the basic definitions and clarifying the graded nature of the fermionic Hilbert–Fock

© Springer Nature Switzerland AG 2019
A. Agarwala, *Excursions in Ill-Condensed Quantum Matter*,
Springer Theses, https://doi.org/10.1007/978-3-030-21511-8_8

space. This allowed us to define the symmetries and identify the unusual symmetries with time reversal, charge conjugation and sublattice. By obtaining the constraints on these symmetries we found the ten classes among which all fermionic Hamiltonians can be classified. Next we obtained the canonical representation of these symmetries which in turn allowed us to construct generic Hamiltonians in all the ten classes. The central results were summarized in Tables 2.1 and 2.2. Our formalism therefore answers the question we had started with. We further implemented this formalism for translationally invariant systems and presented examples of some topological Hamiltonians written in the canonical forms. The insights obtained from this chapter were instrumental in constructing the systems in Chaps. 3 and 4.

Taking a leap, in Chap. 7, we generalized the above symmetry construction for many body Hamiltonians. We organized the Hilbert space in a way that the forms of many-body symmetry operations can be made transparent. We found the generic many-body Hamiltonian structures in each symmetry class. The results from here were summarized in Table 7.1 and Table 7.2. The implication of these results in SYK models, many body localization, random matrix theory were discussed. We also found that some results about a generic system can be obtained just by the structure of the Hamiltonian such as the many body zero modes in AIII class.

However, the classification we have shown is exclusively for a *fermionic* system and the graded structure of the Hilbert space forms the backbone of this formalism (see Fig. 2.1). It will interesting to see how such a formalism can be devised in terms of Majorana fermions and/or bosons. The same graded structure of the vector space is not available in these contexts. Another interesting direction is to see how such a formalism derives for hard-core bosons. Here, the results might be expected to have close semblance to our fermionic results.

In Chap. 3 we used the insights from Chap. 2 and demonstrated topological phases in amorphous systems. The essential motivation was to inquire if a topological phase is at all possible in a system which has no semblance to a lattice? We answered this in the affirmative. Our investigations included demonstration of robust edge states, quantized transport, and a nontrivial topological number—the Bott index. We illustrated similar features in all nontrivial classes in two dimensions and also showed a surface state in three dimensional system. We identified interesting features in the gap closing and the jump of Bott index as the system size is scaled. Importantly, we found that a critical density of the randomly placed lattice sites is required to obtain a topologically nontrivial phase. This work has important implications in the experimental search of topological matter. It opens up the possibility that appropriate density of spin-orbit coupled impurities in an otherwise band gap insulator can make the host topological.

As was discussed in the end of Chap. 3, apart from technological implications, this work opens up important theoretical questions. One—how to derive an analytical theory of the amorphous topological phase which involves no dependence on Brillouin zone concepts. Some of these might go back to underlying framework of c^* algebras used in formulation of a Bott index [1]. There are interesting connections to non-commutative geometry which might also be worth investigating [2]. Another direction is to obtain a deeper understanding of the nature of fermions in the amorphous systems. Dirac fermions in condensed matter systems can occur in two varieties—(i) Wilson and (ii) Susskind–Kogut kind [3]. While the former contains a lone infrared Dirac cone, in the latter, multiple Dirac cones appear, mostly due to lattice symmetries. The amorphous systems we have constructed are manifestations of Wilson fermions. Many standard topological model Hamiltonians such as Haldane model [4], Kane–Mele [5] are forms of the Susskind-Kogut kind. It is important to note that the latter phases are distinct from weak topological phases. It will be interesting to see if there is a sub-classification of topological phases based on this, and that only some of them are realizable in an amorphous setting.

In Chap. 4 we investigated "topological" phases on fractals. Fractal doesn't have integral dimension, neither does it have the notion of bulk and boundary well defined. In order to further maintain the equivalence of all sites, we choose to investigate those fractals where all sites are equally coordinated. Such a system is an outlier to the zoo of topological matter (see [6]). We setup topological Hamiltonians on such a fractal— the same Hamiltonian which produces topologically insulating phases in crystalline lattices. To our surprise we find that we cannot obtain a topologically insulating phase on a fractal. In fact one obtains a metal, however, the current carrying states are chiral. Yet, the Bott index is zero. Moreover, these states can transmit close to unit conductance. The eigenspectrum also has an interesting self-similar spectrum. We also analyzed a higher dimensional fractal object—the Sierpinski tetrahedron, where the surfaces of the fractal are itself fractals! One again finds the system is a metal and the states close to zero energy are corner states. This work shows a new intriguing phase and raises many pertinent questions.

Kitaev's table of topological classification of phases (see Table 1.1) depends crucially on the tenfold symmetry classification of fermions and integer spatial dimension. Where does fractals fall? We can obtain a nontrivial Bott index only when equivalence of all sites is abandoned. How does topological index depend on the nature of boundaries? The principle of bulk boundary correspondence needs to be reinvented for these systems—where the bulk and the boundary cannot be demarcated. Another interesting direction is to analyze other kind of fractals. The fractals which we have analyzed are deterministic fractals—where position of all sites are strongly correlated to each other. Further there is no disorder in this system. Another class of fractals are random or nondeterministic. While these still have a fractal spatial dimension, there is "disorder" in these systems. It will be interesting to see how topological phases appear in such fractals. It might be interesting to separate the effects of—"the loss of dimensionality" and that of "disorder".

In Chap. 5 we turned to analyze another system which hosts a fractal spectrum— the Hofstadter model. Here when one applies a perpendicular magnetic field to a square lattice, the variation of eigenenergies with magnetic flux forms this famous pattern. The question we asked is—how does the pattern change as the bonds of the lattice are randomly removed. We found interesting structures in the decimation of the pattern. We also looked at the effect of such disorder on the transverse conductivity which is a topological quantity. We found that as a function of increasing bond percolation, there is a transition from a Hall phase to an insulating phase, even while the bond percolation value is higher than the classical bond percolation threshold. We also found an interesting filling dependence, where the bands close to band center are least stable to disorder. We contrast parts of our results with those obtained for other kinds of disorder. For future scope, it will be interesting to see the effect of bond percolation on longitudinal conductivity and corroborate it with those from transverse conductivity. It has been numerically found that in the case of Anderson disorder, the variation of transverse and longitudinal conductivity follows a two parameter scaling theory. It will be interesting to see if the effects of bond percolation also follows the same scaling form.

Away from investigations into noninteracting systems, in Chap. 6, we analyzed an interacting model, where a correlated impurity hybridizes with a spin-orbit coupled metal. We found that this can produce emergent fractional spins and a novel Kondo effect. Spin-orbit coupling was shown to produce a bound state, whose energy goes deeper as a function of spin-orbit coupling strength λ as $\lambda^{4/3}$. Interestingly the bound state only has a fraction of the original d-impurity which equals $\frac{2}{3}$. As it turns out, that it is this bound state physics which leads to a realization of a fractional local moment and a corresponding high Kondo temperature. The specific value of 2/3 was shown to be a universal number which only depends on the infrared divergence of the bath density of states. If the divergence is characterized by ϵ^r where $-1 < r < 0$ the fractional moment was found to be $Z = \frac{1}{1-r}$. We established these results using a variety of methods such as mean-field theory, variational calculations and finite temperature quantum Monte Carlo simulations.

Although unreported in this thesis, our investigations in spin-orbit coupled two-dimensional system also produces similar effects. A fractional moment is again formed with a fractional spin of value 2/3. The essential reason being that spin-orbit coupling here also produces an additional divergent contribution to the otherwise constant density of states of the metal. The effect therefore is quite universal. It will be interesting to see if an experimental investigation can detect such a fractional moment. All the models we have investigated are continuum models. It will be worth investigating this physics in a lattice setting and see how the fractional Kondo effect manifests.

Another interesting direction is to understand the implication of this work in the Heavy fermion systems [7, 8]. Heavy fermion materials are strongly correlated systems which have crucial role of f-orbitals. The f-orbital at every atomic site can be considered to be undergoing a "local" Kondo effect. These Kondo effects at every site can undergo hybridization and produce a Kondo insulator. Interesting correlations can arise between various local spins through RKKY interactions which can produce correlated spin phases and rich phase diagrams [8]. However, the wide spectrum of the Heavy fermion physics is built on the underlying Kondo effect—of a unit moment. It will interesting to see how the Heavy fermion physics get modified due to "fractional" moments!

Well, with this, *this thesis* comes to an end.

References

1. Loring TA, Hastings MB (2010) Disordered topological insulators via C*-algebras. Eur Phys Lett 92(6):67004
2. Prodan E, Schulz-Baldes H (2016) Bulk and boundary invariants for complex topological insulators: from K-theory to physics. Springer International Publishing, Berlin
3. Fradkin E (1991) Field theories of condensed matter systems, vol 7. Addison-Wesley, Redwood City
4. Haldane FDM (1988) Model for a quantum hall effect without Landau levels: condensed-matter realization of the "parity anomaly". Phys Rev Lett 61:2015–2018
5. Kane CL, Mele EJ (2005) Z_2 topological order and the quantum spin hall effect. Phys Rev Lett 95:146802
6. Wen XG (2016) Zoo of quantum-topological phases of matter, pp 1–16. arXiv:1610.03911
7. Coleman P (2007) Heavy Fermions: electrons at the edge of magnetism. Wiley Online Library
8. Coleman P (2015) Heavy fermions and the Kondo lattice: a 21st century perspective. ArXiv E-prints, arXiv:1509:05769

Appendix
Methods

A.1 Transport Calculations Using NEGF

In order to calculate the transport characteristics of our random lattices, we model a realistic system by coupling the random lattice to leads and calculating the two terminal conductance(G) using non-equilibrium Green's function (NEGF) method [1]. We first briefly describe the formalism (closely following the presentation in [2]) and then detail the method of implementation in the specific context of the random lattice system. Any essential experimental setup for measuring two-terminal conductance comprises of a device(D) and the connecting leads (left(L) and right(R)). An evaluation of the total transmission for a spinless fermion incident at an energy E, denoted by $T(E)$, can then be quantitatively related to the conductance(G) by $G = \frac{e^2}{h} T(E)$.

The total Hamiltonian for the complete system can be written as,

$$H = \begin{bmatrix} H_L & H_{LD} & 0 \\ H_{DL} & H_D & H_{DR} \\ 0 & H_{RD} & H_R \end{bmatrix} \quad \text{(A.1)}$$

where H_D describes the device, H_L and H_R describes the left and right leads and (H_{DL}, H_{LD}, H_{RD}, H_{DR}) describes the coupling of the device to the leads. While the device is a finite size system, the leads are typically modeled by infinitely large systems of a repeating unit cell. The prescription is to evaluate the self energies due to the right and left lead, $\Sigma_R(E)$ and $\Sigma_L(E)$ and then evaluate the coupling matrices

$$\Gamma_{L(R)}(E) = i\left(\Sigma_{L(R)}(E) - \Sigma_{L(R)}^{\dagger}(E)\right). \quad \text{(A.2)}$$

The Green's function of the device is given by

$$G_D(E) = [E - H_D - \Sigma_L(E) - \Sigma_R(E)]^{-1}. \quad \text{(A.3)}$$

© Springer Nature Switzerland AG 2019
A. Agarwala, *Excursions in Ill-Condensed Quantum Matter*,
Springer Theses, https://doi.org/10.1007/978-3-030-21511-8

The total transmission can be then evaluated using,

$$T(E) = \text{tr}[\Gamma_L(E)G_D^\dagger(E)\Gamma_R(E)G_D(E)]. \tag{A.4}$$

Assuming that H_D is defined on basis of total number $= N_D$ (includes all the sites and associated orbitals(L)), matrices Γ and Σ are $N_D \times N_D$ matrices. Since the leads are semi-infinite, it is not straight forward to calculate their self energies. We now describe the calculation of the self-energy due to the leads.

Calculation of the self energy of the lead: Consider that the leads are made out of a square lattice. This means that a open one-dimensional tight-binding chain (along y) gets repeated in x direction (to negative infinity) for the left lead and to positive infinity for the right lead (see Fig. 3.4). This can be easily generalized for different kinds of leads with appropriate choice of the unit cell. Let the Hamiltonian for this unit cell be given by H_o. For the square lattice example, where the vertical width of the lead has N_y sites, the unit-cell is an open chain of N_y sites. Here an electron can hop from one site to another (in the y-direction) with hopping strength $-t(t = 1)$. H_o is therefore a $N_y \times N_y$ matrix given by

$$H_o = -t \sum_{I=1}^{N_y-1} (c_I^\dagger c_{I+1} + h.c.). \tag{A.5}$$

Now, each of these unit cells are coupled (in the x-direction) to each other through a coupling matrix V. This matrix, in context of the square lattice example, is a diagonal matrix with all diagonal entries($= -1$) (corresponding to the connecting horizontal bonds). This allows us to evaluate the self energy of the leads

$$\Sigma_{l,r}(E) = V \frac{1}{E - H_o - \Sigma_{l,r}(E)} V^\dagger. \tag{A.6}$$

This can be calculated recursively. In order to obtain the stability of the calculation a small complex number η is added to E, which can be progressively reduced as the convergence of the solution improves. Note that $\Sigma_{l,r}$ are $N_y \times N_y$ matrices which are different from $\Sigma_{L,R}$ (defined before) which are matrices having the dimension N_D (that of the device D). One can now obtain the surface part of the Green function $g_{L,R}(E)$

$$g_{L,R}(E) = \frac{1}{E - H_o - \Sigma_{l,r}(E)}. \tag{A.7}$$

Next, $\Sigma_{L(R)}(E)$ can be calculated using,

$$\Sigma_L(E) = H_{DL}g_L(E)H_{LD} \tag{A.8}$$

$$\Sigma_R(E) = H_{DR}g_R(E)H_{RD}. \tag{A.9}$$

Note that $(H_{DL}, H_{LD}, H_{RD}, H_{DR})$ are rectangular matrices where one of the dimensions is N_D and the other side corresponds to the number of sites the device couples to in the respective lead (for the square lattice example=N_y). Having obtained all the essential quantities as required in Eqs. (A.2) and (A.3), one can obtain $T(E)$ as given by Eq. (A.4).

Calculation for the random lattice system: We now describe the method for the random lattice system. Considering density of sites ρ in a box of volume V with L orbitals per site, the Hamiltonian for the device has dimension $N_D = \rho V L$. Depending on the dimension of the system(d), $V \equiv s^d$ where s denotes the length of a single side. We will discuss here the case for $d = 2$. Since defining the edge of a random lattice is not immediately clear, we extend our device by adding one unit cell of the leads to both right and left side of the random lattice (see Fig. 3.4). Since some of the random lattice models we have constructed has two-orbitals per site and some of them have four-orbitals per site, we describe below how the device is extended and how the leads are chosen for each of them.

Class A, D and C: All of these three classes do not have time reversal symmetry and are defined on $L = 2$ orbitals per site. Physically, we can consider $L = 2$ as some atomic orbitals per site for spinless fermions. We connect single orbital square lattice leads to this system. In addition, we append, to both right and left of the random lattice, a $N_y = s$ sized, single orbital tight-binding chain. These appended sites can then be smoothly connected to the semi-infinite leads. The hoppings between these sites of the appended chain and the original orbitals of the random lattice are given by $t(r)$ as described in Eq. (3.5).

Class AII, DIII: Both these systems are $L = 4$ systems. These can be thought of as two copies of class A and class D models, with additional hopping terms (if any) which are allowed by time reversal symmetry. Physically, as is true in the case of topological insulators, these two copies are the spin flavors. We again choose square lattice as the leads of the system, but now with two flavors for both the spins. Each spin square lattice couples to the associated two orbitals in the random lattice through the hopping $t(r)$.

A.2 Calculation of Bott Index

This index is calculated when periodic boundary conditions is applied to the system. The position of any site I (for any orbital α) is given by two coordinates (x_I, y_I). This can be rescaled into two angles on a torus given by (θ_I, ϕ_I) where $0 \leq \theta_I < 2\pi$ and $0 \leq \phi_I < 2\pi$. The next step is to calculate the projector of the occupied states (number of them $\equiv N_o$). We always work at half filling (one fermion per site). Let the occupied single particle states be denoted by $|\psi_j^o\rangle$ ($j = 1, \ldots, N_o$). The projector (P) of the occupied states is given by,

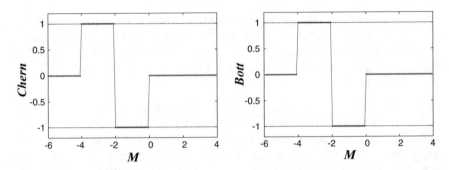

Fig. A.1 **Chern and Bott**: (Left) Chern number for the BHZ model (see Eq. (1.25)) as a function of M ($B = 1$). (Right) Bott index calculated for a 20×20 square lattice with the same Hamiltonian implemented in real space and under periodic boundary conditions

$$P = \sum_{j=1}^{N_o} |\psi_j^o\rangle\langle\psi_j^o|. \tag{A.10}$$

Next, consider two matrices Θ and Φ which are diagonal matrices with elements $\theta_{I\alpha}$ and $\phi_{I\alpha}$ respectively. Note that $\theta_{I\alpha} = \theta_I$ and $\phi_{I\alpha} = \phi_I$. One can now calculate two matrices $U = P\exp(i\Theta)P$ and $W = P\exp(i\Phi)P$. U and W are approximately commuting matrices and the Bott index (m) can be found by evaluating

$$\text{tr}(\log(WUW^\dagger U^\dagger)) = 2\pi i m + r \tag{A.11}$$

where r is a real number. Physically a nontrivial value of Bott index signifies the inability to smoothly deform the occupied wave functions into site localized Wannier orbitals.

A.2.1 Relationship to Chern Number

We illustrate the relationship of the Bott index to Chern number (TKNN invariant [3]) using the example introduced in Chap. 1 (see Eq. (1.25)). This is called the BHZ model for a Chern insulator and is also discussed in [4]. The variation of Chern number for this model was shown in Fig. 1.10 and is reproduced again in Fig. A.1.

Chern number can be calculated by integrating the Berry curvature over the Brillouin zone. We now also calculate the Bott index for a 20×20 square lattice with the same Hamiltonian and under periodic boundary conditions. The result is also shown in Fig. A.1. We can see that both the topological invariants are identical for this crystalline system. A formal mathematical proof of the equivalence of Chern number and Bott index is provided in [5]. Our discussion is intended to provide a physical example of this connection.

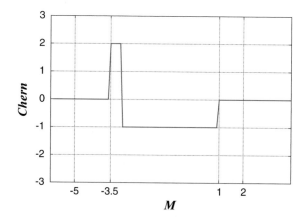

Fig. A.2 BHZ on a triangular lattice: The Chern number as a function of M for the BHZ model when setup on a triangular lattice (see Eq. (A.12))

A.2.2 BHZ Model on a Triangular Lattice

We can implement the class **A** model, as described in Chap. 3, on a triangular lattice. The Hamiltonian is given by

$$H = \sigma_x \left(-\sin(k_x) - \sin\left(\frac{k_x}{2}\right) \cos\left(\frac{\sqrt{3}}{2}k_y\right) \right) + \sigma_y \left(-\sqrt{3}\cos\left(\frac{k_x}{2}\right) \sin\left(\frac{\sqrt{3}}{2}k_y\right) \right)$$
$$+ \sigma_z \left(M + 2 - \left[\cos(k_x) + 2\cos\left(\frac{k_x}{2}\right) \cos\left(\frac{\sqrt{3}}{2}k_y\right) \right] \right). \tag{A.12}$$

The variation of the gap at half-filling as a function of parameter M was shown in Fig. 4.2. The Chern number variation is shown in Fig. A.2. The model therefore provides for a topological system both in square lattice [4] and in a triangular lattice.

References

1. Datta S (1997) Electronic transport in mesoscopic systems. Cambridge University Press, Cambridge
2. Muñoz Rojas F, Jacob D, Fernández-Rossier J, Palacios JJ (2006) Coherent transport in graphene nanoconstrictions. Phys Rev B 74:195417
3. Thouless DJ, Kohmoto M, Nightingale MP, den Nijs M (1982) Quantized hall conductance in a two-dimensional periodic potential. Phys Rev Lett 49:405–408
4. Bernevig BA, Hughes TL (2013) Topological insulators and topological superconductors. Princeton University Press, Princeton
5. Hastings MB, Loring TA (2011) Topological insulators and C-algebras: theory and numerical practice. Ann Phys 326(7):1699–1759

Printed in the United States
By Bookmasters